Discrete Quantum Walks on Graphs and Digraphs

Discrete quantum walks are quantum analogues of classical random walks. They are an important tool in quantum computing and a number of algorithms can be viewed as discrete quantum walks, in particular Grover's search algorithm. These walks are constructed on an underlying graph, and so there is a relation between properties of walks and properties of the graph. This book studies the mathematical problems that arise from this connection, and the different classes of walks that arise. Written at a level suitable for graduate students in mathematics, the only prerequisites are linear algebra and basic graph theory; no prior knowledge of physics is required. The text serves as an introduction to this important and rapidly developing area for mathematicians and as a detailed reference for computer scientists and physicists working on quantum information theory.

CHRIS GODSIL is Distinguished Professor Emeritus at the University of Waterloo. He has written three books: *Algebraic Combinatorics* (1993), *Algebraic Graph Theory* (2004, co-authored with Gordon Royle) and *The Erdos-Ko-Rado Theorem: Algebraic Approaches* (2015, co-authored with Karen Meagher).

HANMENG ZHAN is a postdoctoral fellow at Simon Fraser University. For her thesis on discrete quantum walks via algebraic graph theory, she received two awards from the University of Waterloo: the University Finalist for the Governor General's Gold Medal and the Inaugural Mathematics Doctoral Prize.

London Mathematical Society Lecture Note Series: 484

Discrete Quantum Walks on Graphs and Digraphs

C. GODSIL
University of Waterloo

H. ZHAN
Simon Fraser University

Shaftesbury Road, Cambridge CB2 8EA, United Kingdom

One Liberty Plaza, 20th Floor, New York, NY 10006, USA

477 Williamstown Road, Port Melbourne, VIC 3207, Australia

314–321, 3rd Floor, Plot 3, Splendor Forum, Jasola District Centre, New Delhi – 110025, India

103 Penang Road, #05–06/07, Visioncrest Commercial, Singapore 238467

Cambridge University Press is part of Cambridge University Press & Assessment, a department of the University of Cambridge.

We share the University's mission to contribute to society through the pursuit of education, learning and research at the highest international levels of excellence.

www.cambridge.org
Information on this title: www.cambridge.org/9781009261685
DOI: 10.1017/9781009261692

First published 2023

A catalogue record for this publication is available from the British Library.

ISBN 978-1-009-26168-5 Paperback

Contents

Preface

A discrete quantum walk is determined by a unitary matrix U, the *transition matrix* of the walk. If the initial state of the system is given by a vector z, then the state of the system at time k is $U^k z$. The problem is to choose U and z so that we can do something useful, and indeed we can – Grover showed how an implementation of this setup could be used to enable quantum computers to search a database faster than any known classical algorithm.

The framework we have just described is impossibly general; a quantum computer can conveniently implement only a small subset of the set of unitary matrices. There is also a mathematical difficulty, in that it may be impossible to derive useful predictions of the behaviour of the walk without imposing some structure on U.

As we have described it, the transition matrix U is an operator on the complex inner product space $\mathbb{C}^d$. However, for the reasons just given, much of the work on discrete quantum walks considers the case where U is an operator on the space of complex functions on the arcs (ordered pairs of adjacent vertices) of a graph X. Physically meaningful questions must be expressed in terms of the absolute values of the entries of the powers U^k. Thus, we might ask if, for a given initial state z, there is an integer k such that the absolute values of the entries of U^k are close to being equal.

The goal of our work on this topic has been to attempt to relate the properties of the walk to the properties of the underlying graph, and this book is both an introduction to the topic and a report on our progress.

We start our treatment with the most famous topic, Grover's search algorithm. We offer two approaches, but in both cases we find that the transition matrix arises as a product $U = RC$, where R and C are unitary matrices with simple structure and are defined in terms of an underlying graph. In fact, R and C are both involutions, and the algebra they generate is a matrix representation of the dihedral group. We make use of this fact to determine the spectral decomposition of U in terms of the underlying graph. (If the graph is k-regular on n vertices, U is of order $nk \times nk$, so we have reduced the scale of the problem.) We then apply the resulting theory to the study of properties of our walks, and determine useful parameters. Of course, each time we identify a parameter of a walk, we have introduced a possibly new graph parameter, and many interesting questions raise their heads.

In the second part of the book we relax our assumptions that R and C are involutions. We find that, to properly specify the resulting walks, we must specify a linear ordering on the arcs leaving a vertex. As any graph theorist is aware, embeddings of graphs in an orientable surface are specified by cyclic orderings of the arcs leaving a vertex. Hence we offer a detailed treatment of graph embeddings and graph covers. Following this, we consider walks based on shunts and walks on the line. We close the book with a treatment of what we call vertex-face walks, which are explicitly derived from embeddings of graphs in orientable surfaces.

We note that this book is based on the Ph.D. thesis of the second author (https://uwspace.uwaterloo.ca/handle/10012/13952). The intended audience is mathematicians, particularly those who might be interested in new graph theory problems arising from the study of discrete quantum walks. The book by Portugal [58] provides a complementary view. We do not think any knowledge of physics is required to profit from this work; the required background is linear algebra (spectral decomposition) and some field theory. We have tried to keep things self-contained, but Godsil and Royle [35] may prove a useful backup.

Cambridge University Press has a website devoted to this book at https://www.cambridge.org/gb/academic/subjects/mathematics/discrete-mathematics-information-theory-and-coding/discrete-quantum-walks-graphs-and-digraphs?

1

Grover Search

1.1 States

Any quantum system has a state space, which is a complex inner product space. For us, this will usually be finite dimensional, just $\mathbb{C}^d$ for some d. The actual states are the 1-dimensional subspaces of this vector space. We could specify a subspace U of the complex inner product space V by giving an orthonormal basis $u_1, \ldots, u_k$, but it is often more convenient to define U in terms of the orthogonal projection P onto U – this is the idempotent Hermitian matrix with image equal to U. In fact, if v^* denotes the conjugate transpose of the vector (or matrix) v, then

$$P = \sum_i u_i u_i^*,$$

but, despite appearances, P is independent of the choice of orthonormal basis for U.

Operations on the state space correspond to unitary matrices. If U is unitary and the state of our system is given by a unit vector z, then the vector Uz defines the new state. If we choose to work with projections, our initial state is given by zz^*, and the state after we apply U is Uzz^*U^*.

The outcome of a measurement of a quantum system modelled by $\mathbb{C}^d$ can be taken to be an element of $\{1, \ldots, d\}$. However, the result is actually a random variable: there are probabilities $p_1, \ldots, p_d$ summing to 1, such that we observe outcome i with probability p_i. Thus, we have a probability density defined on the set $\{1, \ldots, d\}$. This means we can view the outcome of a measurement as a probability density. This probability density will depend on the initial state of our system, the operations we apply to the system, and the choice of measurement.

Mathematically, a measurement is represented by a sequence $M_1, \ldots, M_e$ of positive semidefinite matrices such that $\sum_i M_i = I$. The simplest case is when $e = d$ and $M_i = e_i e_i^T$ (here e_i denotes the characteristic vector of i, and ^{T} denotes the transpose). We describe this as 'measurement relative to the standard basis.' If the state of the system is zz^*, then the probability that we observe the ith outcome is

$$\operatorname{tr}(M_i zz^*) = z^* M_i z,$$

which is equal to the inner product $\langle M_i, zz^* \rangle$; if we are measuring relative to the standard basis, the probability is

$$z^* e_i e_i^T z = |\langle z, e_i \rangle|^2.$$

Thus, it is the square of the absolute value of the ith entry of z.

1.2 Discrete Walks

For our purposes, a *discrete quantum walk* is specified by a unitary matrix U. We call it the *transition matrix* of the walk. If U is $d \times d$, we view it as acting on a quantum system with state space $\mathbb{C}^d$. The system evolves under repeated applications of U; thus, if the initial state of the system is represented by the unit vector z, then after m steps, the state of the system would be $U^m z$. If we measure the system after k steps relative to the standard basis, the outcome will be e_j with probability

$$|\langle e_j, U^m z \rangle|^2.$$

Our view of a discrete quantum walk is more general than taken by physicists. We find the generality useful, but there are two problems. The first is mathematical: at this level of generality, we may lack the mathematical tools needed to determine interesting properties of parameters of the walk. The second is physical: some unitary matrices decribe operations that are not easily implemented in practice; thus, we will see that U is usually defined as a product of simple unitary matrices, often sparse.

One common feature of nearly all discrete walks in this book will be that the state space is the set of complex functions on the arcs of a graph. Here an *arc* of a graph is an ordered pair of adjacent vertices. Thus, if X is an undirected graph with m edges, then it has $2m$ arcs, and the associated state space will have dimension $2m$.

1.3 Grover Search

We present one of the most important applications of quantum walks, Grover's search algorithm. Basically we have a system with state space $\mathbb{C}^d$ and two unitary operators R and S. The operators have a special form; they are *reflections*. We explain what this means.

If P is a projection, then

$$(2P - I)^2 = 4P^2 - 4P + I = I,$$

and it follows that $2P - I$ is unitary with order two. It fixes each vector in $\mathrm{im}(P)$ and maps a vector v in $\ker(P) = U^\perp$ to $-v$. Thus, $2P - I$ represents reflection in $\mathrm{im}(P)$.

The simplest case is when $\mathrm{im}(P)$ is 1-dimensional, namely $\mathrm{rk}(P) = 1$. If $\mathrm{im}(P)$ is spanned by a, then

$$P = \frac{1}{\langle a, a \rangle} a a^*$$

and $\ker(P) = a^\perp$. We say that $2P - I$ represents reflection in the hyperplane $a^\perp$.

The operator R is supplied to us and represents reflection in the subspace $e_j^\perp$. We do not know what the value of j is, and we want to determine it. (This is our search problem.) Let $\mathbf{1}$ denote the all-ones vector. The second operator S represents reflection in the orthogonal complement of the vector

$$y = \frac{1}{\sqrt{d}} \mathbf{1}.$$

Grover's strategy is very easy to describe. We initialize our system so that its state is y, we apply the operator $U = RS$ exactly m times, and then we measure relative to the standard basis. If we choose m correctly, the result of the measurement is j, with probability very close to 1. In fact, choosing m to be $O(\sqrt{d})$ will work, beating the classical bound, and this is Grover's algorithm.

In quantum computing there is a standard procedure for encoding 01-valued functions as unitary operators. The operator R is the encoding of a function f that takes the value 1 on j, and is zero on i if $i \neq j$. Clearly, given f we can determine j by trying each input in turn, and on average this will take $\frac{1}{2}d$ tries.

1.4 Justifying Grover's Algorithm

We use a geometric argument to show that Grover's algorithm will work. A real matrix Q represents an orthogonal mapping if $Q^T Q = I$. As

$$1 = \det(QQ^T) = \det(Q)^2,$$

the determinant of an orthogonal mapping is ± 1. A *rotation* is an orthogonal mapping with determinant 1.

Reflections form an important class of orthogonal mappings (which we will be making much use of). If W is a subspace of V, a *reflection in W* is the linear mapping that fixes each element in W and acts as $-I$ on $U^\perp$. Thus, the square of a reflection is the identity, as expected. For our use, the most important case will be reflection in a hyperplane, which can be described as follows. If $a \neq 0$, then the map τ_a defined by

$$\tau_a(x) := x - 2\frac{\langle a, x \rangle}{\langle a, a \rangle} a$$

is reflection in the hyperplane $a^\perp$. It is easy to see that $\tau_a^2 = I$ and $\tau_a(a) = -a$, hence τ_a is a reflection by definition. (You may find it worthwhile to verify that it is an orthogonal mapping.) Since the eigenvalues of τ_a are -1 (with multiplicity of one) and 1 with multiplicity $\dim(V) - 1$, we see that $\det(\tau_a) = -1$.

Now assume that a and b are linearly independent unit vectors with $\cos(\theta) = \langle a, b \rangle$. The product $U = \tau_a \tau_b$ has a determinant of one. Assume that $\dim(V) = n$ and let W be the subspace $a^\perp \cap b^\perp$ of V. Then $\dim(W) = d - 2$ and $W^\perp$ is the 2-dimensional subspace of V spanned by a and b. The restriction of U to W is an orthogonal mapping with determinant 1, and hence it is a rotation.

We claim the restriction of U to $W^\perp$ represents rotation by an angle of 2θ. Since the restriction is a rotation, it suffices to compute the angle between x and Ux for one vector x, and we may take x to be b. Then

$$\tau_a \tau_b(b) = \tau_a(-b) = -b + 2\langle a, b \rangle$$

and so

$$\langle b, Ub \rangle = -1 + 2\langle a, b \rangle^2 = 2\cos(\theta)^2 - 1 = \cos(2\theta).$$

Now we specialize to the case of interest. Assume

$$a := \frac{1}{\sqrt{d}}\mathbf{1}$$

and that b is a standard basis vector. Then

$$\langle a, b \rangle = \frac{1}{\sqrt{d}}$$

and therefore

$$\cos(2\theta) = \frac{2}{d} - 1.$$

Hence, when d is large, U is rotation through an angle a bit less than π, and $-U$ represents a rotation through a small positive angle, ϕ say. As

$$\cos(\phi) \approx 1 - \frac{1}{2}\phi^2,$$

we have

$$\phi \approx \frac{2}{\sqrt{d}}.$$

Accordingly, if

$$N := \left\lfloor \frac{\pi\sqrt{d}}{4} \right\rfloor,$$

then $U^N a$ is very close to b or $-b$. Consequently, the result of a measurement in the standard basis after N applications of U will identify which standard basis vector is equal to b.

1.5 Composite Quantum Systems

A *composite* quantum system is a system whose state space is the tensor product $U \otimes V$, where U and V are the state spaces of two "smaller" quantum systems. A system with state space of this form is said to be *bipartite*. The state space of a system of d qubits is the tensor product of d copies of $\mathbb{C}^2$. We could view this state space as the tensor product of $\mathbb{C}^2$ with $(\mathbb{C}^2)^{\otimes(d-1)}$. A bipartite system models the situation where we have two physicists, traditionally Alice and Bob, each with their own quantum systems. The complete system is described by a tensor product, but Alice and Bob work independently.

Given a bipartite system, we can operate on the individual parts separately; such operations are said to be *local*. More precisely, if R_1 and R_2 are unitary operations on state spaces U_1 and U_2 respectively, then $R_1 \otimes R_2$ is a local unitary operation on $U_1 \otimes U_2$.

Measurements become more complicated, or more interesting, because a measurement carried out on one part is not a measurement on the entire system. If Alice's measurement is specified by positive definite matrices M_r (with $\sum_r M_r = I$) and Bob's by positive semidefinite matrices N_s (with sum $\sum_s N_s = I$), then the Kronecker products

$$M_r \otimes N_s$$

define a measurement on the composite system.

We give an example. Consider the system with state space $\mathbb{C}^n \otimes \mathbb{C}^n$. We think of $\mathbb{C}^n$ as the space of complex functions on the vertices of the complete graph K_n; hence; we may view $\mathbb{C}^n \otimes \mathbb{C}^n$ as the space of complex functions on the arcs of the graph we get by adding a loop to each vertex of K_n. (So $e_u \otimes e_u$ represents a loop on vertex u.)

We introduce three operators on our state space. The first, denoted R, is the permutation operator given by

$$R(e_i \otimes e_j) = e_j \otimes e_i;$$

this is **not** a local operator.

Let τ_j be the operator on $\mathbb{C}^n$ corresponding to reflection about e_j and let τ_1 be reflection in $\mathbf{1}^{\perp}$. Then $\tau_j \otimes I$ and $I \otimes \tau_1$ are local operators.

We note that

$$R(\tau_j \otimes \tau_0)R = \tau_0 \otimes \tau_j$$

and it is not hard to see that, for any integer k,

$$\left(R(\tau_j \otimes \tau_0)\right)^{2k} = (\tau_0 \tau_j)^k \otimes (\tau_j \tau_0)^k.$$

Thus, the action of

$$U := R(\tau_j \otimes \tau_0)$$

on $\mathbb{C}^n \otimes \mathbb{C}^n$ is completely determined by the actions of $\tau_0 \tau_j$ and $\tau_j \tau_0$ on $\mathbb{C}^n$. (We note that $\tau_j \tau_0 = (\tau_0 \tau_j)^{-1}$.)

Since $\tau_0 \tau_j$ is the operator used in Grover's algorithm, it is possible to implement Grover's algorithm using the quantum walk (given by U) on the arcs and loops of K_n. This was first noted by Ambainis, Kempe, and Rivosh [4]. We present the details in the following section.

1.6 Grover via a Quantum Walk on Arcs

Assume $U := R(\tau_j \otimes \tau_0)$, as in the previous section. If we start with the uniform superposition

$$x_0 \otimes x_0 := \frac{1}{n}\mathbf{1} \otimes \mathbf{1},$$

then

$$U^k(x_0 \otimes x_0) \approx e_j \otimes \left((\tau_j \tau_0)^k x_0\right)$$

and measuring the first register at step k (relative to the standard basis) yields e_j with high probability.

If X is a graph, and u and v are two vertices, we write $u{\sim}v$ if u and v are adjacent, equivalently if $\{u, v\}$ is an edge of X. An *arc* is an ordered pair of adjacent vertices, denoted (u, v).

Let X denote the complete graph on n vertices, with one loop on each vertex. (So its adjacency matrix is the all-ones matrix J.) The state space of the above walk is spanned by the characteristic vectors $e_u \otimes e_v$ of the arcs (u, v) of X. Thus, each state can be seen as a complex-valued function on the arcs of X. As an example, the initial state in Grover's search is

$$x_0 \otimes x_0 = \sum_{u \sim v} \frac{1}{n} e_u \otimes e_v,$$

the constant function that maps each arc to $\frac{1}{n}$. Since U acts linearly on $\mathbb{C}^n \otimes \mathbb{C}^n$, it suffices to investigate its effect on the basis

$$\{e_u \otimes e_v : u \sim v\}.$$

The matrix

$$\tau_j \otimes \tau_0 = (2e_j e_j^T - I) \otimes \left(\frac{2}{n}J - I\right)$$

is usually referred to as the *coin operator*, for it acts as if one is flipping a quantum coin to determine which arc to move to, given the current position. Since

$$(\tau_j \otimes \tau_0)(e_u \otimes e_v) = \begin{cases} e_u \otimes \left(\frac{1}{\sqrt{n}}\sum_{w \sim u} e_w\right), & u \neq j, \\ e_u \otimes \left(-\frac{1}{\sqrt{n}}\sum_{w \sim u} e_w\right), & u = j, \end{cases}$$

the result of a coin flip is some superposition of outgoing arcs of current tail u. The matrix R is called the *arc-reversal operator*, as it maps the characteristic vector of (u, v) to the characteristic vector of (v, u). These describe how a quantum walker moves on X: in each step: she flips the coin to redistribute her amplitudes over the outgoing arcs, and then reverses all the arcs she is on.

1.7 Arc-Reversal Grover Walk

Rewrite the unitary matrix of Grover's search as

$$\begin{aligned} U &= R(\tau_j \otimes \tau_0) \\ &= R(I \otimes \tau_0)(\tau_j \otimes I), \end{aligned}$$

and define

$$U_0 := R(I \otimes \tau_0), \quad U_j := \tau_j \otimes I.$$

The first matrix U_0 defines a quantum walk on X, where the coin operator $I \otimes \tau_0$ treats all vertices equally. The second matrix U_j makes a difference between the marked and unmarked vertices: on outgoing arcs of j, it acts as $-I$, while on other arcs it acts as the identity.

The main focus of this book will be quantum walks on graphs with no marked vertices. In this section, we generalize the walk defined by U_0 to an *arc-reversal Grover walk* on any graph; this model was first studied by Watrous [71] and later formalized by Kendon [47].

Let X be a d-regular graph on n vertices. Consider the space $\mathbb{C}^n \otimes \mathbb{C}^d$ spanned by all complex functions on the arcs of X. To each vertex we assign the same *Grover coin*

$$G := \frac{2}{d}J - I.$$

Thus, for vertex u, the amplitude transferred between two outgoing arcs of u is $2/d - 1$ if they are equal, and $2/d$ otherwise. The coin matrix, acting on $\mathbb{C}^n \otimes \mathbb{C}^d$, is then a direct sum of n Grover coins. Since G commutes with all permutations, we can write the coin matrix as $I \otimes G$ under any basis of $\mathbb{C}^n \otimes \mathbb{C}^d$. Let R be the matrix that reverses all arcs, and set

$$U := R(I \otimes G).$$

The quantum walk with U as the transition matrix is an *arc-reversal Grover walk* on X. It is not hard to extend this definition to an irregular graph: simply assign the Grover coin with $d = \deg(u)$ to vertex u.

1.8 Alternative Formulation of Arc-Reversal Walks

A state is a complex function on the arcs of a graph. Hence, it defines a special weighted adjacency matrix W, where the weight W_{uv} (possibly zero) is the amplitude on the arc (u, v), and

$$\sum_{u \sim v} |W_{uv}|^2 = 1.$$

Conversely, given a weighted adjacency matrix W, let $\mathrm{vec}\,(W)$ be the vector obtained from W by concatenating its columns. Clearly, $\mathrm{vec}\,(W)$ is indexed by all pairs of vertices; if we restrict it to the adjacent pairs only, then we have recovered our usual representation of the state.

This motivates an alternative description of arc-reversal walks with Grover coins. Let A be the 01-adjacency matrix of the graph, and D be the diagonal degree matrix. Let $\circ$ denote the Schur or entrywise product of two matrices. For any matrix state W, the arc-reversal operator simply transposes W, and the coin operator sends W to

$$2A((D^{-1}AM) \circ I) - W.$$

In other words, to update the entry W_{uv} after one iteration of the walk, we may first compute the column sum $\langle We_u, \mathbf{1} \rangle$, and then replace W_{uv} by

$$\frac{2}{\deg(u)} \langle We_u, \mathbf{1} \rangle - W_{vu}.$$

Below is a proof for the second statement. In this proof, we use the *vectorization operator* $\mathrm{vec}(\cdot)$, which sends a matrix M to a vector consisting of the columns of M.

1.8.1 Lemma. *Let C be the Grover coin operator, indexed by all pairs of vertices. Then*

$$C \operatorname{vec}(W) = \operatorname{vec} 2A((D^{-1}AM) \circ I) - W.$$

Proof We first write each coin as a weighted adjacency matrix:

$$
\begin{aligned}
C_u &= (AE_{uu}A) \circ \left(\frac{2}{\deg(u)} J - I \right) \\
&= \frac{2}{\deg(u)} AE_{uu}A - \sum_{v \sim u} E_{vv}.
\end{aligned}
$$

Since C is block-diagonal with C_u as the uu-block,

$$
\begin{aligned}
C \operatorname{vec} M &= \left(\sum_u E_{uu} \otimes C_u \right) \operatorname{vec} W \\
&= \sum_u (E_{uu}^T \otimes C_u) \operatorname{vec} W \\
&= \sum_u \operatorname{vec} C_u W E_{uu} \\
&= \operatorname{vec} \sum_u C_u W E_{uu},
\end{aligned}
$$

where the second and third equalities follow from well-known identities of vectorization. Finally, notice that

$$
\begin{aligned}
\sum_u C_u W E_{uu} &= \sum_u \frac{2}{\deg(u)} AE_{uu}AWE_{uu} - \sum_u \sum_{v \sim u} E_{vv} M E_{uu} \\
&= 2A \sum_u \frac{1}{\deg(u)} (e_u^T AWe_u) E_{uu} - \sum_u \sum_{v \sim u} W_{vu} E_{vu} \\
&= 2A((D^{-1}AW) \circ I) - W \circ A \\
&= 2A((D^{-1}AW) \circ I) - W. \qquad \square
\end{aligned}
$$

The matrix $D^{-1}A$ is row stochastic and represents a simple random walk on the graph. This reveals a connection between certain classical walks and quantum walks.

Notes

In general, the coins of a quantum walk do not have to be identical. If we assign $-G$ to a special vertex and G elsewhere, then we have effectively introduced an oracle. This walk was proposed by Ambainis, Kempe, and Rivosh [4] as a quantum algorithm that generalizes Grover's search.

More flexibly, we may assign any $\deg(u) \times \deg(u)$ unitary matrix C_u to a vertex u. However, unless it commutes with all permutations, we will need to specify a linear order on the neighbours of u,

$$f_u : \{1, 2, \cdots, \deg(u)\} \to \{v : u \sim v\},$$

in order to explain what C_u does. Let us refer to the vertex $f_u(j)$ as the jth neighbour of u, and the arc $(u, f_u(j))$ as the jth arc of u. Then, C_u sends the jth arc of u to a superposition of all outgoing arcs of u, in which the amplitudes come from the jth column of C_u:

$$C_u e_j = \sum_{k=1}^{\deg(u)} (e_k^T C_u e_j) e_k.$$

Thus, under the ordering of arcs:

$$\{\{(u, f_u(j)) : j = 1, \cdots, \deg(u)\} : u \in V(X)\},$$

and the transition matrix of our quantum walk is

$$U = R \begin{pmatrix} C_1 & & & \\ & C_2 & & \\ & & \ddots & \\ & & & C_n \end{pmatrix}.$$

The *Fourier coin*

$$F := \frac{1}{\sqrt{d}} (e^{2jk\pi i/d})_{jk}$$

has been frequently studied in the literature. It induces many non-classical behaviors of quantum walks; for example, on the infinite path, the probability distribution is asymmetric about the center [3]. We will visit this model in Chapter 8.

Some coins can be associated with combinatorial structures. If we convert the linear order f_u into a cyclic permutation, then we obtain a *rotation system*, which determines an orientable embedding of a graph (this will be explained in Chapter 6). The readers are invited to show that a unitary circulant matrix commutes with all cyclic permutations if and only if it has simple eigenvalues; this allows us to define, given a fixed $d \times d$ coin, a unique arc-reversal quantum walk for each rotation system of a d-regular graph. In [37], we studied arc-reversal walks on cubic graphs with different rotation systems and found some interesting connections between properties of the walk and properties of the embedding.

2

Two Reflections

There are two major differences between a quantum walk and a classical random walk: first, the evolution is unitary rather than stochastic; second, the transition matrix U does not depend on the graph X only, but also on the coins. In general, little can be said about the relation between U and X. However, the situation for an arc-reversal walk is a bit special, as its transition matrix U is a product of two reflections related to the graph X.

In this chapter, we develop some machinery that applies to any unitary matrix U that can be written as a product of two reflections. A complete characterization of the eigenvalues and eigenspaces of U is given by "lifting" those of a smaller Hermitian matrix constructed from the two reflections. This extends Szegedy's work on direct quantization of Markov chains [65]. The main applications of our machinery are presented in the following chapter and in Chapter 6.

2.1 A Subspace Decomposition

Although our ultimate focus is on products of two reflections, it proves more convenient to work with projections. (We explained the connection between reflections and projections in Section 1.3.)

Let P and Q be two projections acting on $\mathbb{C}^m$. Define

$$U := (2P - I)(2Q - I).$$

Then U lives in the matrix algebra generated by P and Q, denoted $\langle P, Q \rangle$. We will use the following well-known fact to diagonalize U.

2.1.1 Lemma. *Let P and Q be two projections acting on $\mathbb{C}^m$. Then $\mathbb{C}^m$ is a direct sum of 1- and 2-dimensional $\langle P, Q \rangle$-invariant subspaces.*

11

Proof Since P and Q are Hermitian, a subspace of $\mathbb{C}^m$ is $\langle P, Q \rangle$-invariant if and only if its orthogonal complement is $\langle P, Q \rangle$-invariant. Hence $\mathbb{C}^m$ can be decomposed into a direct sum of $\langle P, Q \rangle$-invariant subspaces. Let W be one such subspace.

If $\dim(W) = 1$, then W is spanned by common eigenvectors of P and Q, and we are done. So assume $\dim(W) \geq 2$. Since QPQ is also Hermitian, W is a direct sum of eigenspaces for QPQ. Depending on how QPQ acts on W, we have two cases. Suppose first that QPQ is not zero on W. Then there is $z \in \mathbb{C}^m$ and $\mu \neq 0$ such that

$$QPQz = \mu z.$$

Since

$$\mu Qz = Q(QPQ)z = QPQz = \mu z,$$

the vector z must be an eigenvector for Q as well, so

$$Qz = z,$$

and

$$QPz = QPQz = \mu z.$$

It follows that the subspace spanned by $\{z, Pz\}$ is $\langle P, Q \rangle$-invariant. Now suppose QPQ is zero on W. If Q is also zero on W, then PQ commutes with QP on W, and so W is spanned by common eigenvectors of P and Q. If Q is not zero on W, then it has an eigenvector $z \in W$ with non-zero eigenvalue, that is,

$$Qz = z.$$

Since

$$QPz = QPQz = 0,$$

the subspace spanned by $\{z, Pz\}$ is $\langle P, Q \rangle$-invariant. $\qquad\square$

2.2 Real Eigenvalues

To find the spectral decomposition of U, we first decompose $\mathbb{C}^m$ into a direct sum of 1- and 2-dimensional $\langle P, Q \rangle$-invariant subspaces, and then diagonalize U restricted to each of them. The 1-dimensional $\langle P, Q \rangle$-invariant subspaces are common eigenspaces for P and Q. In fact, they are precisely the eigenspaces for U with real eigenvalues, that is, 1 and -1.

In what follows, we will let $\mathrm{col}(P)$ denote the column space of the matrix P.

2.2.1 Lemma. *Let P and Q be two projections on $\mathbb{C}^m$. Let*

$$U = (2P - I)(2Q - I).$$

The 1-eigenspace for U is the direct sum

$$(\mathrm{col}(P) \cap \mathrm{col}(Q)) \oplus (\ker(P) \cap \ker(Q)),$$

and the (-1)-eigenspace for U is the direct sum

$$(\mathrm{col}(P) \cap \ker(Q)) \oplus (\ker(P) \cap \mathrm{col}(Q)).$$

Proof We prove the first statement. The second statement follows by replacing Q with $I - Q$.

If z is in $\mathrm{col}(P) \cap \mathrm{col}(Q)$, then $Pz = z$ and $Qz = z$, so

$$Uy = (2P - I)(2Q - I)y = y.$$

If z is in $\ker(P) \cap \ker(Q)$, then $Pz = 0$ and $Qz = 0$, so

$$Uz = (2P - I)(2Q - I)y = -(-y) = y.$$

By linearity, every vector in

$$(\mathrm{col}(P) \cap \mathrm{col}(Q)) \oplus (\ker(P) \cap \ker(Q))$$

is an eigenvector for U with eigenvalue 1. Now suppose $Uz = z$ for some $z \in \mathbb{C}^m$. Then

$$(2Q - I)z = (2P - I)z.$$

Thus, $Pz = Qz$ and $(I - P)z = (I - Q)z$. From the decomposition

$$z = Pz + (I - P)z,$$

we see that z lies in

$$(\mathrm{col}(P) \cap \mathrm{col}(Q)) \oplus (\ker(P) \cap \ker(Q)). \qquad \square$$

2.3 Complex Eigenvalues

It remains to construct eigenvectors for U with non-real eigenvalues. As indicated in the proof of Lemma 2.1.1, the eigenspaces of PQP play a crucial rule in providing the 2-dimensional U-invariant subspaces. In practice, we will work with the eigenspaces of a smaller Hermitian matrix that is related to PQP, as we describe now.

Let L be a matrix whose columns form an an orthonormal basis for $\mathrm{im}(Q)$. Then $L^*L = I$, and so LL^* represents orthogonal projection onto $\mathrm{im}\,Q$. Therefore, $Q = LL^*$. Note that

$$QPQz = \mu z$$

if and only if

$$L^*PL(L^*z) = \mu(L^*z).$$

Consequently, for any $\mu \neq 0$, the map $z \mapsto L^*z$ is an isomorphism from the μ-eigenspace of QPQ to the μ-eigenspace of L^*PL, with inverse given by $y \mapsto Ly$. We claim that the eigenspaces for L^*PL with non-zero eigenvalues provide all eigenvectors for U with non-real eigenvalues. Our proof uses the following standard result on eigenvalue interlacing; for a reference, see Horn and Johnson [41, Ch 4].

2.3.1 Theorem. *Let A be a Hermitian matrix. Let L be a matrix with $L^*L = I$. If $B = L^*AL$, the eigenvalues of B interlace those of A.* $\square$

2.3.2 Lemma. *Let P and Q be projections on $\mathbb{C}^m$. Let*

$$U = (2P - I)(2Q - I).$$

Suppose $Q = LL^*$ for some matrix L with orthonormal columns. The eigenvalues of L^*PL lie in $[0, 1]$. Let y be an eigenvector for L^*PL. Let $z = Ly$. We have the following correspondence between eigenvectors for L^*PL and eigenvectors for U.

  (i) *If y is an eigenvector for L^*PL with eigenvalue 1, then*

$$z \in \mathrm{col}(P) \cap \mathrm{col}(Q).$$

  (ii) *If y is an eigenvector for L^*PL with eigenvalue 0, then*

$$z \in \ker(P) \cap \mathrm{col}(Q).$$

  (iii) *If y is an eigenvector for L^*PL with eigenvalue $\mu \in (0, 1)$, and $\theta \in \mathbb{R}$ satisfies that $2\mu - 1 = \cos(\theta)$, then*

$$(\cos(\theta) + 1)z - (e^{i\theta} + 1)Pz$$

is an eigenvector for U with eigenvalue $e^{i\theta}$, and

$$(\cos(\theta) + 1)z - (e^{i\theta} + 1)Pz$$

is an eigenvector for U with eigenvalue $e^{-i\theta}$.

Proof Since the columns of L are orthonormal, the eigenvalues of L^*PL interlace those of P, which are 0 and 1. If

$$L^*PLy = y,$$

then

$$yL^*(I - P)Ly = 0,$$

and it follows from the positive-definiteness of $I-P$ that $Ly \in \mathrm{col}(P)$. Similarly, if

$$L^*PLy = 0,$$

then $Ly \in \ker(P)$.

Finally, suppose

$$L^*PLy = \mu y$$

for some $\mu \in (0, 1)$. Then the subspace spanned by $\{z, Pz\}$ is U-invariant:

$$U \begin{pmatrix} z & Pz \end{pmatrix} = \begin{pmatrix} z & Pz \end{pmatrix} \begin{pmatrix} -1 & -2\mu \\ 2 & 4\mu - 1 \end{pmatrix}.$$

To find linear combinations of z and Pz that are eigenvectors of U, we diagonalize the matrix

$$\begin{pmatrix} -1 & -2\mu \\ 2 & 4\mu - 1 \end{pmatrix}.$$

It has two eigenvalues: $e^{i\theta}$ with eigenvector

$$\begin{pmatrix} -\cos(\theta) - 1 \\ e^{i\theta} + 1 \end{pmatrix},$$

and $e^{-i\theta}$ with eigenvector

$$\begin{pmatrix} -\cos(\theta) - 1 \\ e^{-i\theta} + 1 \end{pmatrix}.$$

Since $0 < \mu < 1$, these two eigenvalues are distinct, and

$$\frac{\cos(\theta) + 1}{e^{\pm i\theta} + 1}I - P$$

is invertible, so

$$(\cos(\theta) + 1)z - (e^{\pm i\theta} + 1)Pz$$

is indeed an eigenvector for U with eigenvalue $e^{\pm i\theta}$. $\qquad\square$

The preceding construction preserves orthogonality – eigenvectors for U obtained from orthogonal eigenvectors for L^*PL are also orthogonal. In the following section, we summarize information on all eigenspaces for U we have seen so far, including their multiplicities. As a consequence, their direct sum is precisely $\mathbb{C}^m$.

2.4 Multiplicities

Let

$$P = KK^*$$

for some matrix K with orthonormal columns. Define

$$S := L^*K.$$

This matrix largely determines the spectrum of U.

2.4.1 Lemma. *Let P and Q be projections on $\mathbb{C}^m$. Let*

$$U = (2P - I)(2Q - I).$$

The 1-eigenspace of U is the direct sum

$$(\mathrm{col}(P) \cap \mathrm{col}(Q)) \oplus (\ker(P) \cap \ker(Q)),$$

which has dimension

$$m - \mathrm{rk}(P) - \mathrm{rk}(Q) + 2\dim(\mathrm{col}(P) \cap \mathrm{col}(Q)).$$

Moreover, if $K^*K = L^*L = I$ and

$$P = KK^*, \quad Q = LL^*$$

and

$$S = L^*K,$$

then the map $y \mapsto Ly$ is an isomorphism from the 1-eigenspace of SS^* to $\mathrm{col}(P) \cap \mathrm{col}(Q)$.

Proof For the multiplicity, note that

$$\dim(\ker(P) \cap \ker(Q)) = \dim\left(\ker\begin{pmatrix} P \\ Q \end{pmatrix}\right)$$

$$= m - \mathrm{rk}\begin{pmatrix} P & Q \end{pmatrix}$$

$$= m - \dim\left(\mathrm{col}\begin{pmatrix} P & Q \end{pmatrix}\right)$$

$$= m - \dim(\mathrm{col}(P) + \mathrm{col}(Q))$$

$$= m - (\mathrm{rk}(P) + \mathrm{rk}(Q) - \dim(\mathrm{col}(P) \cap \mathrm{col}(Q))).$$

The isomorphism follows from Lemma 2.3.2 and the previous discussion. $\square$

2.4.2 Lemma. *Let P and Q be projections on $\mathbb{C}^m$, with $K^*K = L^*L = I$ and*

$$P = KK^*, \quad Q = LL^*.$$

Let

$$S = L^*K.$$

Let

$$U = (2P - I)(2Q - I).$$

The (-1)-eigenspace of U is the direct sum

$$(\mathrm{col}(P) \cap \ker(Q)) \oplus (\ker(P) \cap \mathrm{col}(Q)),$$

which has dimension

$$\mathrm{rk}(P) + \mathrm{rk}(Q) - 2\,\mathrm{rk}(S).$$

Moreover, the map $y \mapsto Ky$ is an isomorphism from $\ker(S)$ to $\mathrm{col}(P) \cap \ker(Q)$, and the map $y \mapsto Ly$ is an isomorphism from $\ker(S^*)$ to $\ker(P) \cap \mathrm{col}(Q)$.

Proof We prove the last part of the statement, from which the dimension follows. If

$$Sy = 0,$$

then

$$QKy = LSy = 0.$$

Hence

$$Ky \in \mathrm{col}(P) \cap \ker(Q).$$

Further, since K has full column rank, this map is injective. On the other hand, for any $z \in \mathrm{col}(P) \cap \ker(Q)$, there is some y such that

$$z = Ky$$

and

$$0 = Qz = LSy = L^*LSy = Sy,$$

which implies that

$$y \in \ker(S).$$

The argument for the second linear map is similar. $\square$

2.4.3 Lemma. *Let P and Q be projections on $\mathbb{C}^m$ such that*

$$P = KK^*, \quad Q = LL^*$$

with $K^*K = L^*L = I$. Let

$$S = L^*K.$$

Let

$$U = (2P - I)(2Q - I).$$

The dimensions of the eigenspaces for U with non-real eigenvalues sum to

$$2\,\mathrm{rk}(S) - 2\,\dim(\mathrm{col}(P) \cap \mathrm{col}(Q)).$$

Let $\mu \in (0, 1)$ be an eigenvalue of SS^*. Let θ be such that $\cos(\theta) = 2\mu - 1$. The map

$$y \mapsto ((\cos(\theta) + 1)I - (e^{i\theta} + 1)P)Ly$$

is an isomorphism from the μ-eigenspace of SS^* to the $e^{i\theta}$-eigenspace of U, and the map

$$y \mapsto ((\cos(\theta) + 1)I - (e^{-i\theta} + 1)P)Ly$$

is an isomorphism from the μ-eigenspace of SS^* to the $e^{-i\theta}$-eigenspace of U.

Proof By Lemma 2.3.2 and Lemma 2.4.1, the eigenspaces for SS^* with eigenvalues in $(0, 1)$ provide

$$2(\mathrm{rk}(SS^*) - \dim(\mathrm{col}(P) \cap \mathrm{col}(Q)))$$

orthogonal eigenvectors for U. Combining this with Lemma 2.4.2, we see that they span the orthogonal complement of the (± 1)-eigenspaces. The isomophisms now follow from Lemma 2.3.2. $\qquad\square$

For normalization purposes, note that

$$\|((\cos(\theta) + 1) - (e^{\pm i\theta} + 1)P)Ly\|^2 = \sin^2(\theta)(\cos(\theta) + 1)\|y\|^2.$$

This will become useful when we compute the orthogonal projection onto the $e^{\pm i\theta}$-eigenspace.

Notes

With the theory developed in this chapter, we can derive the spectral decomposition of any matrix in the algebra generated by two projections P and Q. By comparison, Szegedy [65] computed the eigenvalues and eigenvectors specifically for the matrix $(2P - I)(2Q - I)$.

For most known models of discrete quantum walks, the transition matrix is a product of two reflections $U = (2P - I)(2Q - I)$. Examples include the staggered walks [58, Chapter 8], where P and Q are the characteristic matrices of partitions of the vertex set into cliques; the bipartite walks, where P and Q are the characteristic matrices of partitions of the edge set of a bipartite graph; the walks using gems (Ch. 6), where P and Q arise from an embedding of a graph; and many other walks, which are discussed in [59].

3

Applications

Here we present applications of the machinery we developed in the previous chapter. We show that the spectrum of the transition matrix for the arc-reversal walk on a graph X is determined by the spectrum of X. We construct examples of arc-reversal walks that admit perfect state transfer and we explain the connection with Szegedy's model. Finally, we offer a second proof that Grover's algorithm works.

3.1 Graph Spectra versus Walk Spectra

Let X be a connected d-regular graph on n vertices, and U the transition matrix of the arc-reversal walk on X. In this section, we show that the spectrum of X determines the spectrum of U. More specifically, eigenvalues of X provide the real parts of eigenvalues of U, and eigenvectors of X can be lifted to eigenvectors of U by two incidence matrices.

Recall that

$$U = R(I \otimes G),$$

where R is the arc-reversal matrix, and G the $d \times d$ Grover coin. Since

$$R^2 = (I \otimes G)^2 = I,$$

all observations in the previous section apply. To see what R and $I \otimes G$ reflect about, we introduce four incidence matrices: the tail-arc incidence matrix D_t, the head-arc incidence matrix D_h, the arc-edge incidence matrix M, and the vertex-edge incidence matrix B.

The tail-arc incidence matrix D_t and the head-arc incidence matrix D_h are matrices with rows indexed by the vertices, and columns by the arcs. If u is a vertex and a is an arc, then $(D_t)_{u,a} = 1$ if u is the initial vertex of a, and $(D_h)_{u,a} = 1$ if a ends on u, and 0 otherwise.

20

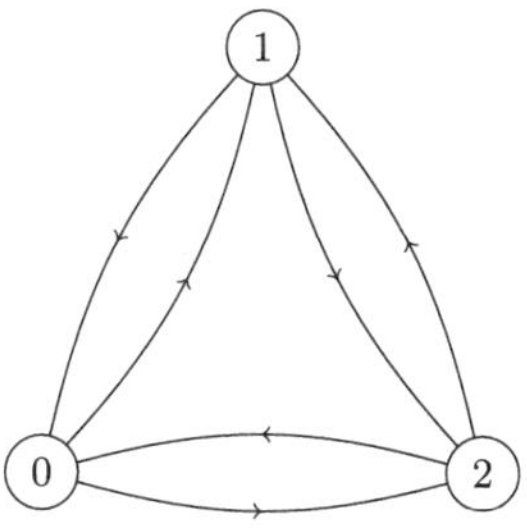

Figure 3.1 K_3

The arc-edge incidence matrix M is a matrix with rows indexed by the arcs and columns by the edges. If a is an arc and e is an edge, then $M_{a,e} = 1$ if a is one direction of e, and 0 otherwise.

The vertex-edge incidence matrix B is a matrix with rows indexed by the vertices and columns by the edges. If u is a vertex and e is an edge, then $B_{u,e} = 1$ if u is one endpoint of e, and 0 otherwise.

As an example, the following are the four incidence matrices associated with K_3 (see Figure 3.1) with vertices $\{0, 1, 2\}$.

$$
D_t = \begin{array}{c} \\ 0 \\ 1 \\ 2 \end{array}
\begin{array}{cccccc}
(0,1) & (0,2) & (1,0) & (1,2) & (2,0) & (2,1) \\
1 & 1 & 0 & 0 & 0 & 0 \\
0 & 0 & 1 & 1 & 0 & 0 \\
0 & 0 & 0 & 0 & 1 & 1
\end{array}
$$

$$
D_h = \begin{array}{c} \\ 0 \\ 1 \\ 2 \end{array}
\begin{array}{cccccc}
(0,1) & (0,2) & (1,0) & (1,2) & (2,0) & (2,1) \\
0 & 0 & 1 & 0 & 1 & 0 \\
1 & 0 & 0 & 0 & 0 & 1 \\
0 & 1 & 0 & 1 & 0 & 0
\end{array}
$$

$$
M = \begin{array}{c} \\ (0,1) \\ (0,2) \\ (1,0) \\ (1,2) \\ (2,0) \\ (2,1) \end{array}
\begin{array}{ccc}
\{0,1\} & \{0,2\} & \{1,2\} \\
1 & 0 & 0 \\
0 & 1 & 0 \\
1 & 0 & 0 \\
0 & 0 & 1 \\
0 & 1 & 0 \\
0 & 0 & 1
\end{array}
$$

$$
B = \begin{array}{c} \\ 0 \\ 1 \\ 2 \end{array}
\begin{array}{ccc} \{0,1\} & \{0,2\} & \{1,2\} \\ \left(\begin{array}{ccc} 1 & 1 & 0 \\ 1 & 0 & 1 \\ 0 & 1 & 1 \end{array} \right) \end{array}
$$

Next, we list some useful identities satisfied by these incidence matrices.

3.1.1 Lemma. *Let X be a d-regular graph. Let A be the adjacency matrix of X. Let D_t and D_h be the tail-arc incidence matrix and the head-arc incidence matrix, respectively. Let M be the arc-edge incidence matrix. Let B be the vertex-edge incidence matrix. Let G be the $d \times d$ Grover coin. The following identities hold:*

   (i) $D_t D_t^T = D_h D_h^T = dI$.
  (ii) $M^T M = 2I$.
 (iii) $D_t D_h^T = D_h D_t^T = A$.
 (iv) $BB^T = A + dI$.
  (v) $D_t M = D_h M = B$.
 (vi) $D_t R = D_h$.
(vii) $R = MM^T - I$.
(viii) $I \otimes G = \frac{2}{d} D_t^T D_t - I \otimes I$.

Proof We give a proof for (iii). Let u and v be two vertices of X. We have

$$
\begin{aligned}
(D_t D_h^T)_{uv} &= \langle D_t^T e_u, D_h^T e_v \rangle \\
&= |\{(a,b) : \{a,b\} \in E(X), a = u, b = v\}| \\
&= \begin{cases} 1, & \{u,v\} \in E(X) \\ 0, & \{u,v\} \notin E(X). \end{cases}
\end{aligned}
$$

Therefore, $D_t D_h^T = A$. Since A is symmetric, we also have $D_h D_t^T = A$. The remaining identities can be verified in a similar manner. $\qquad\square$

As a consequence, R is a reflection about $\mathrm{col}(M)$, while $I \otimes G$ is a reflection about $\mathrm{col}(D_t^T)$. We now prove the spectral relation between U and A. The following theorem shows that all eigenspaces of U with non-real eigenvalues are completely determined by the eigenspaces of X with eigenvalues in $(-d, d)$. It also gives a concrete description of how to 'lift' eigenvalues and eigenvectors of X to those of U.

3.1.2 Theorem. *Let X be a d-regular graph. Let D_t and D_h be the tail-arc incidence matrix and the head-arc incidence matrix, respectively. Let U be the* *transition matrix of the arc-reversal Grover walk on X. The multiplicities of the non-real eigenvalues of U sum to $2n - 4$ if X is bipartite, and $2n - 2$ otherwise. Let y be an eigenvector for X with eigenvalue $\lambda \in (-d, d)$. Let $\theta \in \mathbb{R}$ be such that $\lambda = d\cos(\theta)$. Then*

$$D_t^T y - e^{i\theta} D_h^T y$$

is an eigenvector for U with eigenvalue $e^{i\theta}$, and

$$D_t^T y - e^{-i\theta} D_h^T y$$

is an eigenvector for U with eigenvalue $e^{-i\theta}$.

Proof Let

$$K := \frac{1}{\sqrt{2}} M, \quad L := \frac{1}{\sqrt{d}} D_t^T, \quad S := L^* K.$$

Let B be the vertex-edge incidence matrix of X. According to Lemma 2.4.3, the eigenspaces for U with non-real eigenvalues are determined by eigenspaces for

$$SS^* = \frac{1}{2d} BB^T = \frac{1}{2d}(A + dI).$$

Let

$$\mu := \frac{\lambda + d}{2d}.$$

Then $0 < \mu < 1$ and $2\mu - 1 = \cos(\theta)$. Moreover,

$$Ay = \lambda y$$

if and only if

$$SS^* y = \mu y.$$

Thus, using identities in Lemma 3.1.1, we obtain two eigenvectors for U as stated. $\qquad\square$

After normalization, we obtain the eigenprojections for non-real eigenvalues of U.

3.1.3 Corollary. *Let X be a d-regular graph. Let D_t and D_h be the tail-arc incidence matrix and the head-arc incidence matrix, respectively. Let U be the transition matrix of the arc-reversal Grover walk on X. Let λ be an eigenvalue of X that is neither d nor $-d$. Let E_λ be the orthogonal projection onto* the λ-eigenspace of X. Suppose $\lambda = d\cos(\theta)$ for some $\theta \in \mathbb{R}$. Then the $e^{i\theta}$-eigenprojection of U is

$$\frac{1}{2d\sin^2(\theta)}(D_t - e^{i\theta}D_h)^T E_\lambda (D_t - e^{-i\theta}D_h),$$

and the $e^{-i\theta}$-eigenprojection of U is

$$\frac{1}{2d\sin^2(\theta)}(D_t - e^{-i\theta}D_h)^T E_\lambda (D_t - e^{i\theta}D_h). \qquad \square$$

We also characterize the (± 1)-eigenspaces of U. In particular, their multiplicities depend on parameters of X.

3.1.4 Lemma. *Let X be a d-regular graph. Let D_t be the tail-arc incidence matrix. Let M be the arc-edge incidence matrix. Let U be the transition matrix of the arc-reversal Grover walk on X. The 1-eigenspace of U is*

$$(\mathrm{col}(M) \cap \mathrm{col}(D_t^T)) \oplus (\ker(M^T) \cap \ker(D_t))$$

with dimension

$$\frac{nd}{2} - n + 2.$$

Moreover, the projection onto $\mathrm{col}(M) \cap \mathrm{col}(D_t^T)$ is given by

$$\frac{1}{d}D_t^T E_d D_t = \frac{1}{nd}J,$$

where E_d is the projection onto the d-eigenspace of X.

Proof By Lemma 2.4.1, the 1-eigenspace is the direct sum:

$$(\mathrm{col}(M) \cap \mathrm{col}(D_t^T)) \oplus (\ker(M^T) \cap \ker(D_t)),$$

where

$$\mathrm{col}(M) \cap \mathrm{col}(D_t^T) = D_t\,\mathrm{col}(E_d).$$

Note that $\mathrm{col}(D_t^T)$ consists of vectors that are constant over the outgoing arcs of each vertex, and $\mathrm{col}(M)$ consists of vectors that are constant over each pair of opposite arcs. Since X is connected,

$$\mathrm{col}(M) \cap \mathrm{col}(D_t^T) = \mathrm{span}\{\mathbf{1}\}.$$

The multiplicity follows from the fact that $\mathrm{rk}(M) = nd/2$ and $\mathrm{rk}(D_t) = n$. $\quad\square$

3.1.5 Lemma. *Let X be a d-regular graph. Let D_t be the tail-arc incidence matrix. Let M be the arc-edge incidence matrix. Let B be the vertex-edge incidence matrix. Let U be the transition matrix of the arc-reversal Grover walk on X. If X is bipartite, the (-1)-eigenspace of U is*

$$M \ker(B) \oplus D_t^T \ker(B^T)$$

with dimension

$$\frac{nd}{2} - n + 2.$$

Moreover, the projection onto $D_t^T \ker(B^T)$ is given by

$$\frac{1}{d} D_t^T E_{-d} D_t,$$

where E_{-d} is the projection onto the $(-d)$-eigenspace of X. If X is not bipartite, the (-1)-eigenspace of U is

$$M \ker(B),$$

with dimension

$$\frac{nd}{2} - n.$$

Proof By Lemma 2.4.2, the (-1)-eigenspace of U is

$$M \ker(B) \oplus D_t^T \ker(B^T),$$

where

$$\ker(B^T) = \mathrm{col}(E_{-d}).$$

Note that $\mathrm{rk}(B) = n - 1$ if X is bipartite, and $\mathrm{rk}(B) = n$ otherwise. $\qquad\square$

The spectra of variants of U are also of interest for producing graph isomorphism algorithms. In [23, 24], Emms, Severini, Wilson, and Hancock proposed a scheme to distinguish non-isomorphic graphs, based on the spectrum of the positive support of U^3. Godsil and Guo [31] then studied the relation between the spectra of positive supports of U, U^2, and U^3 in greater detail. Later in [32], Godsil, Guo, and Myklebust found two non-isomorphic strongly regular graphs whose positive supports of U^3 have the same spectrum.

3.2 Perfect State Transfer

Quantum walks were shown to be universal for quantum computation [16, 53, 67]. An important ingredient in implementing the universal quantum gates using quantum walks is perfect state transfer. Loosely speaking, a graph admits perfect state transfer from vertex u to vertex v if, for some real number t,

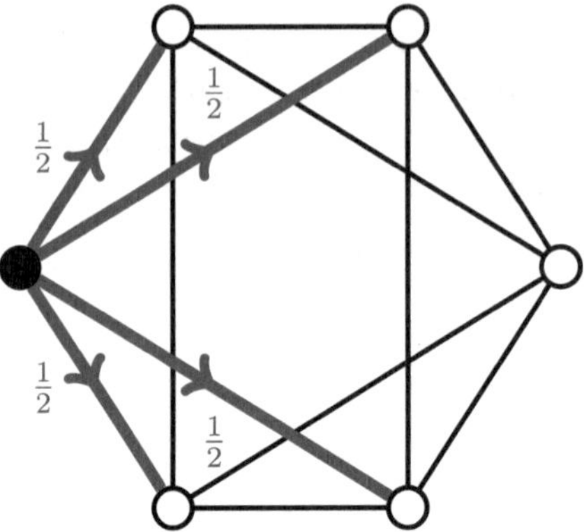

Figure 3.2 Initial state

measuring the system at step t yields vertex v with certainty, given that the system 'concentrated' on vertex u at the beginning. For discrete quantum walks, this is equivalent to requiring the initial state to be a superposition over the outgoing arcs of u, and the final state to be a superposition over the outgoing arcs of v. Sometimes there are more restrictions on the initial and final states; we will give a formal definition later.

In the following four sections, we derive necessary and sufficient conditions for perfect state transfer to occur, and provide an infinite family of circulant graphs that admit antipodal perfect state transfer.

Let X be a d-regular graph on n vertices. An arc-reversal Grover walk takes place in $\mathbb{C}^n \otimes \mathbb{C}^d$. Suppose we start with a uniform superposition over the outgoing arcs of u, which is a special state that concentrates on u (see Figure 3.2):

$$\frac{1}{\sqrt{d}} e_u \otimes \mathbf{1}.$$

If there is a unit vector $x \in \mathbb{C}^d$ such that

$$U^k \left(\frac{1}{\sqrt{d}} e_u \otimes \mathbf{1} \right) = e_v \otimes x,$$

then we say X admits *perfect state transfer* from u to v if $u \neq v$, and X is *periodic* at u if $u = v$. While this definition does not impose further conditions on the final state, in the arc-reversal Grover walk, the only possible choice of x is

$$\frac{1}{\sqrt{d}} \mathbf{1},$$

(see Figure 3.3) as we show now.

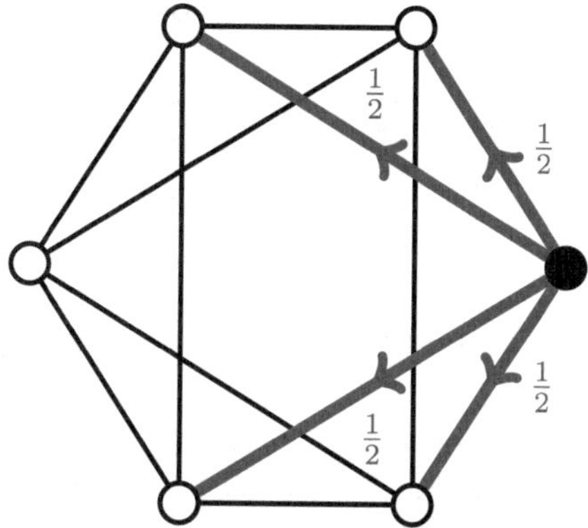

Figure 3.3 Final state

3.2.1 Lemma. *Let X be a regular graph. Let U be the transition matrix of the arc-reversal Grover walk on X. If X admits perfect state transfer from u to v at time k, then*

$$U^k \left(\frac{1}{\sqrt{d}} e_u \otimes \mathbf{1} \right) = \frac{1}{\sqrt{d}} e_v \otimes \mathbf{1}.$$

Proof Suppose

$$U^k \left(\frac{1}{\sqrt{d}} e_u \otimes \mathbf{1} \right) = e_v \otimes x.$$

Since U has real entries, all entries in x are real. Moreover, as $\mathbf{1} \otimes \mathbf{1}$ is an eigenvector for U with eigenvalue 1,

$$\left\langle \mathbf{1} \otimes \mathbf{1}, \frac{1}{\sqrt{d}} e_u \otimes \mathbf{1} \right\rangle = \left\langle \mathbf{1} \otimes \mathbf{1}, U^k \left(\frac{1}{\sqrt{d}} e_u \otimes \mathbf{1} \right) \right\rangle = \langle \mathbf{1} \otimes \mathbf{1}, e_v \otimes x \rangle.$$

If X is d-regular, then it follows that

$$\langle \mathbf{1}, x \rangle = \sqrt{d}.$$

On the other hand, by Cauchy–Schwarz,

$$|\langle \mathbf{1}, x \rangle| \leq \|\mathbf{1}\| \|x\| = \sqrt{d},$$

with equality held if and only if x is a scalar multiple of $\mathbf{1}$. Therefore, x must be equal to $\mathbf{1}$. $\square$

3.3 Characterization of Perfect State Transfer

From previous discussion, we notice that if perfect state transfer occurs, then both the initial state and the final state lie in $\mathrm{col}(D_t^T)$. Thus, an equivalent definition for perfect state transfer from u to v at time k is

$$U^k D_t^T e_u = D_t^T e_v.$$

Our characterization of perfect state transfer relies heavily on this observation.

3.3.1 Lemma. *Let X be a d-regular graph. Let U be the transition matrix of the arc-reversal Grover walk on X. Let $\lambda = d\cos(\theta)$ be an eigenvalue of X that is neither d nor $-d$. Let E_λ be the projection onto the λ-eigenspace of X, and let $F_\pm$ be the projection onto the $e^{\pm i\theta}$-eigenspace of U. Then*

$$D_t F_\pm D_t^T = \frac{d}{2} E_\lambda.$$

Proof By Lemma 3.1.3,

$$
\begin{aligned}
2d\sin^2(\theta) D_t F_\pm D_t^T &= D_t(D_t - e^{\pm i\theta} D_h)^T E_\lambda (D_t - e^{\mp i\theta} D_h) \\
&= (dI - e^{\pm i\theta} A) E_\lambda (dI - e^{\mp i\theta} A) \\
&= d^2 |1 - e^{i\theta}\cos(\theta)|^2 E_\lambda \\
&= d^2 \sin^2(\theta) E_\lambda. \qquad\qquad \square
\end{aligned}
$$

3.3.2 Theorem. *Let X be a d-regular graph, with spectral decomposition*

$$A = \sum_\lambda \lambda E_\lambda.$$

Then the arc-reversal Grover walk on X admits perfect state transfer from u to v at time k if and only if all of the following hold.

(i) For each λ, we have $E_\lambda e_u = \pm E_\lambda e_v$.

(ii) If $E_\lambda e_u = E_\lambda e_v \neq 0$, then there is an even integer j such that

$$\lambda = d\cos(j\pi/k).$$

(iii) If $E_\lambda e_u = -E_\lambda e_v \neq 0$, then there is an odd integer j such that

$$\lambda = d\cos(j\pi/k).$$

Proof Let U be the transition matrix of the arc-reversal Grover walk on X. Consider the spectral decomposition of U:

$$U = \sum_r e^{i\theta_r} F_r.$$

There is perfect state transfer from u to v at time k if and only if

$$\sum_r e^{ik\theta_r} F_r D_t^T e_u = D_t^T e_v,$$

or equivalently, for each r,

$$e^{ik\theta_r} F_r D_t^T e_u = F_r D_t^T e_v. \tag{3.3.1}$$

We prove that Equation (3.3.1) holds if and only if (i), (ii), and (iii) hold. Depending on r, there are three cases.

Suppose $e^{i\theta_r} = 1$. Equation (3.3.1) says that

$$F_r D_t^T e_u = F_r D_t^T e_v.$$

By Lemma 3.1.4, this holds if and only if

$$\frac{1}{nd} J e_u = D_t^T E_d e_u = D_t^T E_d e_v = \frac{1}{nd} J e_v \neq 0,$$

if and only if

$$E_d e_u = E_d e_v \neq 0.$$

Clearly $d = d\cos(0)$, which satisfies (ii).

Suppose $e^{i\theta_r} = -1$. By Lemma 3.1.5,

$$F_r D_t^T = \frac{1}{d} D_t^T E_{-d} D_t D_t^T = D_t^T E_d.$$

Thus, Equation (3.3.1) holds if and only if

$$(-1)^k F_r D_t^T e_u = F_r D_t^T e_v,$$

that is,

$$(-1)^k D_t^T E_{-d} e_u = D_t^T E_{-d} e_v.$$

If X is not bipartite, then $E_{-d} = 0$ and

$$F_r D_t^T e_u = F_r D_t^T e_v = 0.$$

Otherwise,

$$E_{-d} e_u = E_{-d} e_v \neq 0$$

if u and v are in the same colour class, and

$$E_{-d} e_u = -E_{-d} e_v \neq 0$$

if they are in different colour classes. Clearly

$$-d = d\cos\left(\frac{k\pi}{k}\right),$$

which satisfies (i) and (ii).

Finally suppose $e^{i\theta_r} \neq \pm 1$. Equation (3.3.1) says that

$$e^{ik\theta_r} F_r D_t^T e_u = F_r D_t^T e_v.$$

By Lemma 3.3.1,

$$D_t F_r D_t^T = \frac{d}{2} E_\lambda,$$

so

$$\frac{de^{ik\theta_r}}{2}(E_\lambda)_{uu} = e^{ik\theta_r}\langle F_r D_t^T e_u, D_t^T e_u\rangle$$

$$= \langle F_r D_t^T e_v, D_t^T e_u\rangle = \frac{d}{2}(E_\lambda)_{uv}$$

$$\in \mathbb{R}.$$

Therefore, Equation (3.3.1) holds if and only if one of the following occurs:

(a) $E_\lambda e_u = E_\lambda e_v = 0$;
(b) $E_\lambda e_u = E_\lambda e_v \neq 0$, and $e^{ik\theta_r} = 1$;
(c) $E_\lambda e_u = -E_\lambda e_v \neq 0$, and $e^{ik\theta_r} = -1$. $\square$

The three conditions in Theorem 3.3.2 are symmetric in u and v. As a consequence, perfect state transfer is symmetric in the initial and final states, and it implies periodicity at both vertices.

3.3.3 Corollary. *Let X be a regular graph. Consider the arc-reversal Grover walk on X. If there is perfect state transfer from u to v at time k, then there is perfect state transfer from v to u at time k, and X is periodic at both u and v at time $2k$.* $\square$

3.4 Strongly Cospectral Vertices

Let X be a graph with spectral decomposition

$$A = \sum_\lambda \lambda E_\lambda.$$

In [27], the first author of this book defined the *eigenvalue support* of a vertex u to be the set

$$\{\lambda : E_\lambda e_u \neq 0\}.$$

Let $\phi(t)$ be the characteristic polynomial of X, and $\phi_u(t)$ the characteristic polynomial of the vertex-deleted subgraph $X \backslash u$. Then

$$((tI - A)^{-1})_{a,a} = \sum_r \frac{e_u^T E_r e_a}{t - \theta_r}$$

and it follows that the eigenvalue support of u consists of roots of the following polynomial:

$$\psi_u(t) := \frac{\phi(t)}{\gcd(\phi(t), \phi_u(t))}.$$

(The machinery in play here is discussed at greater length in Sections 8.12 and 8.13 of [35].) We see that Theorem 3.3.2 thus gives necessary and sufficient conditions on $\psi_u(t)$ for X to be periodic at u.

3.4.1 Theorem. *Suppose $\psi_u(t)$ has degree ℓ. Then vertex u is periodic at time k if and only if the polynomial*

$$z^\ell \psi_u \left(\frac{d}{2} \left(z + \frac{1}{z} \right) \right)$$

is a factor of $z^k - 1$.

Proof Setting $u = v$ in Theorem 3.3.2, we see that u is periodic at time k if and only if each eigenvalue λ in the eigenvalue support of u is of the form

$$\lambda = \frac{d}{2}(e^{j\pi i/k} + e^{-j\pi i/k}),$$

for some even integer j, or equivalently,

$$z^\ell \psi_u \left(\frac{d}{2} \left(z + \frac{1}{z} \right) \right)$$

divides $z^k - 1$. $\qquad\qquad\square$

One open problem is to characterize graphs with periodic vertices. Although the condition in Theorem 3.4.1 is local on the eigenvalue support of a vertex, it is satisfied when the entire graph is periodic, that is, when all eigenvalues of the graph are d times the real parts of some kth roots of unity. Hence, it is useful to study graphs for which

$$z^n \phi \left(\frac{d}{2} \left(z + \frac{1}{z} \right) \right)$$

is a factor of $z^k - 1$. In [74], Yoshie investigated periodic arc-reversal Grover walks on distance-regular graphs and found all Hamming graphs and Johnson graphs that are periodic.

Two vertices u and v in X are *cospectral* if the vertex-deleted subgraphs $X \backslash u$ and $X \backslash v$ have the same characteristic polynomial, that is,

$$\phi_u(t) = \phi_v(t).$$

We say two vertices u and v are *strongly cospectral* if

$$E_\lambda e_u = \pm E_\lambda e_v$$

for each eigenvalue λ of X. Strong cospectrality has been thoroughly studied by Godsil and Smith [36]; we cite a useful characterization in what follows.

3.4.2 Theorem. *Let X be a graph with spectral decomposition*

$$A = \sum_\lambda \lambda E_\lambda.$$

Two vertices u and v in X are strongly cospectral if and only if both

(i) u and v are cospectral; and
(ii) for every eigenvalue λ of X, the vectors $E_\lambda e_u$ and $E_\lambda e_v$ are parallel. $\qquad\square$

3.5 An Infinite Family

Conditions (ii) and (iii) in Theorem 3.3.2 lead us to consider regular graphs whose eigenvalues are given by real parts of $2k$th roots of unity. A *circulant graph* $X = X(\mathbb{Z}_n, \{g_1, \ldots, g_d\})$ is a Cayley graph over $\mathbb{Z}_n$ with the inverse-closed connection set

$$\{g_1, \ldots, g_d\} \subseteq \mathbb{Z}_n.$$

If ψ is a character of $\mathbb{Z}_n$, then ψ is also an eigenvector for X with eigenvalue

$$\psi(g_1) + \cdots + \psi(g_d).$$

Note that this is a sum of real parts of nth roots of unity. We show that circulant graphs whose connection sets satisfy a simple condition admit perfect state transfer. The following can be found in Zhan [75].

3.5.1 Theorem. *Let ℓ be an odd integer. For any distinct integers a and b such that $a + b = \ell$, the arc-reversal Grover walk on the circulant graph $X(\mathbb{Z}_{2\ell}, \{a, b, -a, -b\})$ admits perfect state transfer at time 2ℓ from vertex 0 to vertex ℓ.*

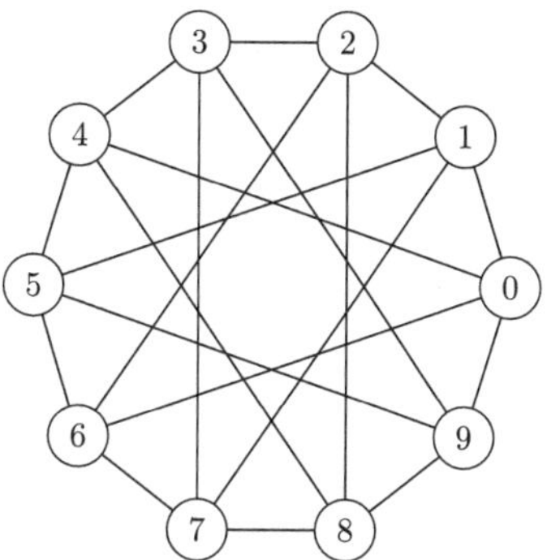

Figure 3.4 $X(\mathbb{Z}_{10}, \{1, 4, 6, 9\})$

Proof The eigenvalues of X are

$$\lambda_j = e^{aj\pi/\ell} + e^{-aj\pi/\ell} + e^{bj\pi/\ell} + e^{-bj\pi/\ell}$$

$$= 2\cos\left(\frac{aj\pi}{\ell}\right) + 2\cos\left(\frac{bj\pi}{\ell}\right),$$

for $j = 0, 1, \ldots, 2n - 1$. Since ℓ is odd and $a + b = \ell$, when j is odd,

$$\lambda_j = 0,$$

and when j is even,

$$\lambda_j = 4\cos\left(\frac{2aj\pi}{2\ell}\right).$$

It suffices to check the parity condition in Theorem 3.3.2 for each eigenvector of X. Since $a + b = \ell$, vertex u and $u + \ell$ have the same neighbours, so

$$A(e_u - e_{u+\ell}) = 0.$$

We see from the multiplicity of 0 that for $u = 0, 1, \ldots, \ell - 1$, the vectors $e_u - e_{u+\ell}$ form an orthogonal basis for $\ker(A)$. Thus $y_u = -y_v$ if y is an eigenvector for X with eigenvalue 0, and $y_u = y_v$ if y is any other eigenvector for X. $\qquad\square$

Figures 3.4 and 3.5 show two examples with perfect state transfer.

3.6 Other Coins

The previous sections were devoted to arc-reversal Grover walks. In the literature, other coins have been studied as well, such as the Fourier coin:

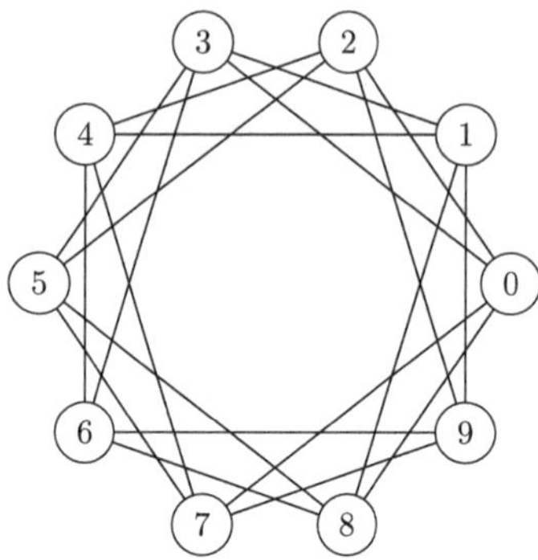

Figure 3.5 $X(\mathbb{Z}_{10}, \{2, 3, 7, 8\})$

$$F := \frac{1}{\sqrt{d}}(e^{2jk\pi i/d})_{jk}.$$

Note that $F^4 = I$, so the techniques in Section 2 do not apply. However, for graphs with special structures, one can still study the arc-reversal Fourier walk analytically. In [49], Krovi and Brun computed the hitting time of an arc-reversal walk on the hypercube Q_d and showed that, for some initial state, the hitting time relative to

$$U = R(I \otimes F)$$

could be infinite. This is in sharp contrast to the polynomial hitting time relative to

$$U = R(I \otimes G),$$

as proved by Kempe [46]. In what follows we give another example showing how coins may affect the behaviour of a quantum walk.

3.6.1 Theorem. *Let $X = K_{m,n}$. For each vertex u, let f_u be a linear order on its neighbours. Suppose $f_u = f_v$ whenever u and v are in the same colour class. Let C_n be an $m \times m$ unitary coin of order k, and attach it to each vertex of degree n. Let C_m be an $n \times n$ unitary coin of order ℓ, and attach it to each vertex of degree m. If U is the transition matrix of the arc-reversal walk on X with coins C_m and C_n, then*

$$U^{2\mathrm{lcm}(k,\ell)} = I.$$

Proof Up to permuting the row and the columns, the transition matrix can be written as

$$U = R \begin{pmatrix} C_n & & & & & & \\ & \ddots & & & & & \\ & & C_n & & & & \\ & & & C_m & & & \\ & & & & \ddots & & \\ & & & & & C_m \end{pmatrix}.$$

Let E_{ij} be the 01-matrix whose ij-entry is 1. Then

$$R = E_{12} \otimes \left(\sum_{i=1}^{m} \sum_{j=1}^{n} E_{ji} \otimes E_{ij} \right) + E_{21} \otimes \left(\sum_{i=1}^{m} \sum_{j=1}^{n} E_{ij} \otimes E_{ji} \right).$$

Thus,

$$U = R(E_{11} \otimes I_m \otimes C_n + E_{22} \otimes I_n \otimes C_m)$$

$$= E_{12} \otimes \left(\sum_{i=1}^{m} \sum_{j=1}^{n} E_{ji} \otimes E_{ij} C_m \right) + E_{21} \otimes \left(\sum_{i=1}^{m} \sum_{j=1}^{n} E_{ij} \otimes E_{ji} C_n \right).$$

Therefore,

$$U^2 = E_{11} \otimes \left(\sum_{j=1}^{n} \sum_{t=1}^{m} E_{jt}(C_m)_{jt} \right) \otimes C_n + E_{22} \otimes \left(\sum_{i=1}^{m} \sum_{s=1}^{n} E_{is}(C_n)_{is} \right) \otimes C_m$$

$$= \begin{pmatrix} C_m \otimes C_n & 0 \\ 0 & C_n \otimes C_m \end{pmatrix}.$$

It follows that the order of U^2 divides the order of $C_m \otimes C_n$, that is, $\mathrm{lcm}(k, m)$.

$\square$

3.7 Szegedy's Model

In the previous sections, we thoroughly studied the spectral decomposition of an arc-reversal walk in order to give exact analysis on properties such as perfect state transfer. Effectively, the arc-reversal model is a special case of Szegedy's quantization of Markov chains [65], as we explain now.

Let X be a graph with n vertices. Let M be the transition matrix of a Markov chain on X. The quantized walk takes place on $\mathbb{C}^n \otimes \mathbb{C}^n$: we will construct two reflections τ_t and τ_h, both of size $n^2 \otimes n^2$, and set

$$U = \tau_t \tau_h.$$

First, take the square root of every entry in M, and let N denote the resulting matrix. Next, define two partitions of the ordered pairs:

$$Q_h = \big(e_1 \otimes (Ne_1) \quad \cdots \quad e_n \otimes (Ne_n)\big),$$
$$Q_t = \big((N^T e_1) \otimes e_1 \quad \cdots \quad (N^T e_n) \otimes e_n\big).$$

Since M is doubly stochastic,

$$Q_t^T Q_t = Q_h^T Q_h = I.$$

In other words, if we view X as a digraph with arcs $V(X) \times V(X)$ and let π_t and π_h be the partitions of the arcs based on their tails and heads, respectively, then Q_t and Q_h are weighted characteristic matrices of π_t and π_h, with weights determined by the Markov chain. Thus,

$$\tau_t = 2Q_t Q_t^T - I,$$
$$\tau_h = 2Q_h Q_h^T - I$$

are two reflections about $\mathrm{col}(Q_t)$ and $\mathrm{col}(Q_h)$, respectively. The quantum walk determined by $U = \tau_t \tau_h$ is called the *direct quantization of M*.

The arc-reversal model is closely related to the direct quantization of the simple random walk. For instance, take the following Markov chain on K_3:

$$M = \begin{array}{c} \\ 1 \\ 2 \\ 3 \end{array} \begin{array}{ccc} 1 & 2 & 3 \end{array} \left(\begin{array}{ccc} 0 & 1/2 & 1/2 \\ 1/2 & 0 & 1/2 \\ 1/2 & 1/2 & 0 \end{array}\right).$$

Then we have

$$Q_h = \begin{array}{c} \\ 11 \\ 12 \\ 13 \\ 21 \\ 22 \\ 23 \\ 31 \\ 32 \\ 33 \end{array} \begin{array}{ccc} 1 \quad & 2 \quad & 3 \end{array} \left(\begin{array}{ccc} 0 & 0 & 0 \\ 1/\sqrt{2} & 0 & 0 \\ 1/\sqrt{2} & 0 & 0 \\ 0 & 1/\sqrt{2} & 0 \\ 0 & 0 & 0 \\ 0 & 1/\sqrt{2} & 0 \\ 0 & 0 & 1/\sqrt{2} \\ 0 & 0 & 1/\sqrt{2} \\ 0 & 0 & 0 \end{array}\right), \quad Q_t = \begin{array}{c} \\ 11 \\ 12 \\ 13 \\ 21 \\ 22 \\ 23 \\ 31 \\ 32 \\ 33 \end{array} \begin{array}{ccc} 1 \quad & 2 \quad & 3 \end{array} \left(\begin{array}{ccc} 0 & 0 & 0 \\ 0 & 1/\sqrt{2} & 0 \\ 0 & 0 & 1/\sqrt{2} \\ 1/\sqrt{2} & 0 & 0 \\ 0 & 0 & 0 \\ 0 & 0 & 1/\sqrt{2} \\ 1/\sqrt{2} & 0 & 0 \\ 0 & 1/\sqrt{2} & 0 \\ 0 & 0 & 0 \end{array}\right).$$

Note that despite the presence of the loops at vertex 1, 2 and 3, the corresponding rows in Q_h and Q_t are all zero, so

$$U = (2Q_t Q_t^T - I)(2Q_h Q_h^T - I)$$

acts as zero on the subspace spanned $e_{(1,1)}$, $e_{(2,2)}$ and $e_{(3,3)}$. Meanwhile, the restriction of U to the orthogonal complement is precisely

$$\left(\frac{2}{3}D_t D_t^T - I\right)\left(\frac{2}{3}D_h D_h^T - I\right) = R\left(\frac{2}{3}D_h D_h^T - I\right) R\left(\frac{2}{3}D_h D_h^T - I\right),$$

where R is the arc-reversal matrix defined as before. Hence, one step of the direct quantization of M is equivalent to two steps of the arc-reversal walk on K_3.

3.8 Justifying Grover's Algorithm, Again

We have seen an geometric proof of Grover's algorithm. In this section, we provide an algebraic proof, using the machinery in Chapter 2.

Recall the unitary operator in Grover's algorithm:

$$U = RS,$$

where R is the reflection about $\mathrm{col}(\mathbf{1})$, and S is the reflection about $\mathrm{col}(e_j)$. Let

$$L = \frac{1}{\sqrt{d}}\mathbf{1}, \quad K = e_j,$$

and

$$P = KK^*, \quad Q = LL^*.$$

Then

$$U = (2P - I)(2Q - I).$$

From Chapter 2, we see that the spectral decomposition of U is largely determined by

$$S = L^*K = \frac{1}{\sqrt{d}}.$$

More specifically, there is only one real eigenvalue, that is, -1, with multiplicity $d - 2$ and eigenspace

$$\ker(P) \cap \ker(Q) = e_j^\perp \cap \mathbf{1}^\perp.$$

There are two conjugate complex eigenvalues $e^{\pm i\theta}$ with

$$\cos(\theta) = \frac{2}{d} - 1,$$

and their eigenspaces are given by

$$\mathrm{col}\left(\frac{2}{\sqrt{d}}\mathbf{1} - (e^{\pm i\theta} + 1)e_j\right).$$

Since the (-1)-eigenspace is orthogonal to the initial state, by the spectral decomposition,

$$U^N\mathbf{1} = \alpha\mathbf{1} + \beta e_j,$$

where α is a multiple of

$$e^{iN\theta} + e^{-iN\theta} = 2\cos(N\theta).$$

If d is large, then $\cos(\theta) \approx -1$, and so

$$\cos(\theta) \approx \frac{1}{2}(\pi - \theta)^2 - 1.$$

Thus, when

$$N = \left\lfloor \frac{\pi\sqrt{d}}{4} \right\rfloor,$$

we have

$$\cos(N\theta) = \pm\cos(N(\pi - \theta)) \approx 0.$$

3.9 Element Distinctness

In this section, we briefly discuss another well-known quantum algorithm, due to Ambainis [2], for the element distinctness problem. Here we are given n elements $\{1, 2, \cdots, n\}$ and a binary function $f : \{1, 2, \cdots, n\} \to \{0, 1\}$. Our task is to determine if there is a pair (x, y) such that $f(x) = f(y)$, and, if so, find (x, y).

We may think of the element distinctness problem as searching for some marked vertex on a graph. There are various ways to do this, but to minimize the evaluations of f, the graph we choose will be a Johnson graph $J(n, k)$, and the marked vertices will be k-subsets that contain a pair (x, y) with $f(x) = f(y)$. As two vertices (k-subsets) in a Johnson graph are adjacent if they differ in exactly one element, each time we move to an adjacent vertex, we only have to evaluate f once, and this is how the oracle complexity of the algorithms ties to the number of steps we run a quantum walk on $J(n, k)$.

Just like Grover's search algorithm, we will initialize our system to the constant vector on the arcs of $J(n, k)$, and run the quantum walk with transition matrix

$$U = RCO$$

a number of steps. Ambainis showed that, with appropriately chosen k, the quantum walk will find a marked vertex using $O(n^{2/3})$ evaluations of f.

3.10 Arc-Reversal Walks with Weighted Grover Coins

The Grover coins prove to be effective for search purposes - in a quantum walk search algorithm, the system evolves from a uniform state to a state that almost concentrates on a target vertex. Sometimes, one may wish to see a different, possibly opposite phenomenon on the quantum walk. For example, can the system *reach* a probability distribution that is uniform over all vertices, regardless of the initial state? This may be achieved by arc-reversal walks as well, using some variants of Grover coins.

Consider a weighted adjacency matrix W, not necessarily real or symmetric, of a graph X. The ab-entry of W is the weight assigned to the arc (a, b), if such arc exists, and zero otherwise. We say W is *quantized* if

$$(W \circ \overline{W})\mathbf{1} = \mathbf{1}.$$

There is a cheap way to construct such W through normalized adjacency matrix $\widehat{A}$: simply let W_{ab} be the square root of $\widehat{A}_{ab}$; we then have

$$W = \frac{1}{\sqrt{k}}A$$

if X is k-regular.

A quantized adjacency matrix W naturally defines a weighted tail-arc incidence matrix N_t, where

$$(N_t)_{u,(a,b)} = \begin{cases} W_{ab}, & u = a, \\ 0, & u \neq 1. \end{cases}$$

In the simplest case, $W_{ab} = \sqrt{\widehat{A}_{ab}}$, and so N_t coincides with the normalized tail-arc incidence matrix $\widehat{D}_t$. Thus,

$$C = 2N_t^* N_t - I$$

can be seen as a generalization the Grover coin matrix. We will call each block of C a *weighted Grover coin*. The *Hadamard coin*,

$$H = \frac{1}{\sqrt{2}}\begin{pmatrix} 1 & 1 \\ 1 & -1 \end{pmatrix},$$

is a special case, as

$$H = \frac{1}{2 - \sqrt{2}}\begin{pmatrix} 1 \\ \sqrt{2} - 1 \end{pmatrix}\begin{pmatrix} 1 & \sqrt{2} - 1 \end{pmatrix} - I.$$

The next chapter deals with the average state of a quantum walk. For arc-reversal walks with Grover coins, the average probability distribution is unlikely to be uniform over the vertices. However, with appropriately chosen weights, this is possible. In fact, if X is a Cayley graph over an abelian group, then almost all weightings that respect the connection set will do.

Notes

The techniques we use are very similar to those employed in continuous quantum walks. For an introduction to this topic, see Section 4.6. A thorough treatment of continuous-time perfect state transfer can be found in Coutinho's Ph.D. thesis [18].

While there have been numerous results on perfect state transfer in continuous quantum walks [6, 7, 8, 15, 19, 20, 21, 43, 44, 48], less is known on the discrete side, as the extra coins make it harder to analyze the transition operator. Most of the examples in discrete quantum walks were sporadic, and there was no infinite family of k-regular graphs with perfect state transfer, for any $k \geq 3$. Kurzynski and Wojcik [50] showed that perfect state transfer on cycles can be achieved in discrete quantum walks. In their paper, they also discussed how to convert the position dependence of couplings into the position dependence of coins. Barr, Proctor, Allen, and Kendon [10] investigated discrete quantum walks on variants of cycles, and found some families that admit perfect state transfer with appropriately chosen coins and initial states. In [73], Yalcnkaya and Gedik proposed a scheme to achieve perfect state transfer on paths and cycles using a recovery operator. With various setting of coin flippings, Xiang Zhan et al [76] also showed that an arbitrary unknown two-qubit state can be perfectly transferred in 1-dimensional or 2-dimensional lattices. Recently, Stefanak and Skoupy analyzed perfect state transfer in perturbed quantum walks on stars [63] and complete bipartite graphs [64] between marked vertices: in $K_{n,n}$, perfect state transfer occurs between any two marked vertices, while in $K_{m,n}$ with $m \neq n$, perfect state transfer only occurs between two marked vertices on the same side.

The type of perfect state transfer we considered has a special form-the initial state lives in $\mathrm{col}(D_t^T)$. Hence, the final state also lies in $\mathrm{col}(D_t^T)$, and we are able to characterize such phenomenon using graph spectra. In theory, for any unit vector x,

$$e_u \otimes x$$

could serve as the initial state that concentrates on u. We do not know whether perfect state transfer can happen if x does not lie in $\mathrm{col}(C)$.

4

Averaging

In the previous chapter, our quantum walker discovered a simple rule to move on a d-regular graph: at each step, she pushes her part on arc (u, v) towards arc (v, w), with "relocation amplitude" $2/d - 1$ if w and v are equal, and $2/d$ otherwise. After exploring for a while, she starts to modify the rule. First, the arc that receives relocation amplitude $2/d - 1$ does not have to be the inverse of the previous one – she could pick the special arc in her own way. Second, these amplitudes do not have to be real – she could toss any complex coin as long as it stays unitary. Finally, the underlying graph does not have to be regular or undirected – she could assign different coins to different vertices based on their outdegrees. However, as time goes, she notices some common phenomena of these quantum walks, due to the nature of unitarity.

The aim of this chapter is to study the limiting behavior of a quantum walk while assuming as little as possible. To allow this level of generality, we suppose the underlying graph X is directed, and U is simply a unitary matrix indexed by the arcs of X. We will consider the applications to specific quantum walks in later chapters.

We start by describing the evolution of a quantum walk in the density matrix formalism, as it cleans up the discussion on various forms of probabilities. Following this, we show that while the instantaneous probability distribution of a quantum walk does not converge, its Cesàro sum does exist, and can be expressed using the spectral idempotents of U. We then explore how fast the time-averaged probability distribution converges to this limit. In particular, four upper bounds on the mixing time are given, with tightness determined by our knowledge of the quantum walk. For the limiting distribution itself, we study a matrix that encodes the limiting probabilities over the arcs, called the average mixing matrix. We prove that it is flat, that is, all entries are of constant modulus, if and only if U has simple eigenvalues with flat eigenprojections; this is a useful characterization, as flat average mixing matrix guarantees uniform limiting distribution over the vertices, regardless of the initial state. Finally, we extend some results on the average mixing matrix in continuous quantum walks to discrete quantum walks. The majority of this chapter comes from a paper [37] by the authors of this book.

4.1 Positive Semidefinite Matrices

We review some of the theory of positive semidefinite matrices. A matrix M is *positive semidefinite* if it is Hermitian and $x^*Mx \geq 0$ for all vectors x. The identity matrix is positive semidefinite; more generally, if P is a projection, then

$$x^*Px = x^*P^2x = x^*P^*Px = \|Px\|^2$$

and so it is positive semidefinite. We write $M \succeq N$ to denote that $M - N$ is positive semidefinite. A matrix is *positive definite* if it is positive semidefinite and invertible.

4.1.1 Theorem. *If M is a Hermitian matrix, the following are equivalent:*

(a) $M \succeq 0$.
(b) The eigenvalues of M are non-negative.
(c) $M = C^*C$ for some matrix C.

We can strengthen (c): if $M \succeq 0$, there is a positive semidefinite matrix N such that $M = N^2$. It also follows from (c) that if $M \succeq 0$, then $A^*MA \succeq 0$ and, from this, it follows that if N is a principal submatrix of M and $M \succeq 0$, then $N \succeq 0$ – in particular the diagonal entries of M are non-negative.

The set of all $d \times d$ positive semidefinite matrices is a closed convex cone. Thus, if $M, N \succeq 0$, then $M + N \succeq 0$ and, more generally, any convex combination of positive semidefinite matrices is positive semidefinite. The positive semidefinite matrices with trace one form a compact convex set; its extreme points are the positive semidefinite matrices with rank one. This implies that any positive semidefinite matrix is a convex combination of positive semidefinite matrices with rank one.

A matrix D is a *density matrix* if $D \succeq 0$ and $\mathrm{tr}(D) = 1$. A density matrix with rank one is called a *pure state*; an arbitrary density matrix can be expressed as a convex combination of pure states, but this expression is far from being unique. We note that both states and measurements involve positive semidefinite matrices.

We define an inner product on the space of $d \times d$ matrices by

$$\langle A, B \rangle := \mathrm{tr}(A^* B).$$

A matrix M is positive semidefinite if and only $\langle M, N \rangle \geq 0$ for each positive semidefinite matrix N. (The convex cone of positive semidefinite matrices is self dual.) We note that

$$\langle zz^*, M \rangle = z^* M z.$$

If $M, N \succcurlyeq 0$, then $\langle M, N \rangle = 0$ if and only if $MN = 0$. Hence $z^* M z = 0$ if and only if $Mz = 0$.

If $M, N \succcurlyeq 0$, then $M \otimes N \succcurlyeq 0$.

4.2 Density Matrices

Let X be a digraph with m arcs. A discrete quantum walk on X is determined by some unitary transition matrix U acting on $\mathbb{C}^m$. Given initial state x_0, at step k, the system is in state

$$x_k := U^k x_0.$$

If we perform a measurement in the standard basis, then the quantum walker is found on arc a with probability

$$P_{x_0, a}(k) := |\langle e_a, U^k x_0 \rangle|^2.$$

We may express the right-hand side using the trace inner product, that is,

$$|\langle e_a, U^k x_0 \rangle|^2 = e_a^T U^k x_0 x_0^* (U^k)^* e_0 = \langle U^k x_0 x_0^* (U^k)^*, e_a e_a^T \rangle.$$

Note that both $U_k x_0 x_0^* (U^k)^*$ and $e_a e_a^T$ are positive semidefinite matrices with trace one. This motivates us to describe quantum walks in a different way, using density matrices.

A *density matrix* is a positive semidefinite matrix ρ with $\mathrm{tr}(\rho) = 1$. All $m \times m$ density matrices form a convex set, with extreme points being the rank one projections, that is, $\rho = xx^*$ for some unit vector $x \in \mathbb{C}^m$. Thus, there is a one-to-one correspondence between the extreme points and the quantum states we have seen; these states are called *pure states*. The remaining density matrices represent probabilistic ensembles of pure states, also called *mixed states*. For example, if one is uncertain about the system state in $\mathbb{C}^2$, but knows that it is e_1 with probability 50% and e_2 with probability 50%, then the density matrix is

$$\frac{1}{2} \begin{pmatrix} 1 & 0 \\ 0 & 1 \end{pmatrix} = \frac{1}{2} e_1 e_1^T + \frac{1}{2} e_2 e_2^T.$$

However, we also have

$$\frac{1}{2}\begin{pmatrix} 1 & 0 \\ 0 & 1 \end{pmatrix} = \frac{1}{2}\begin{pmatrix} \frac{1}{\sqrt{2}} \\ \frac{1}{\sqrt{2}} \end{pmatrix}\begin{pmatrix} \frac{1}{\sqrt{2}} & \frac{1}{\sqrt{2}} \end{pmatrix} + \frac{1}{2}\begin{pmatrix} \frac{1}{\sqrt{2}} \\ -\frac{1}{\sqrt{2}} \end{pmatrix}\begin{pmatrix} \frac{1}{\sqrt{2}} & -\frac{1}{\sqrt{2}} \end{pmatrix},$$

so a density matrix does not necessarily determine the probabilistic ensemble of pure states. For more discussion on pure states and mixed states, see [11, 45, 55].

Now let's revisit the quantum walk on X. Suppose we start with a pure state, say $\rho_0 = x_0 x_0^*$. At step k, the system is in state

$$\rho_k := U^k \rho_0 (U^k)^*.$$

If we perform a measurement in the standard basis, then the system collapses to state $e_a e_a^T$ with probability

$$P_{\rho_0,a}(k) = \langle \rho_k, e_a e_a^T \rangle,$$

that is, the inner product of the pre-measurement state and post-measurement state.

As a special case, when $\rho_0 = e_b e_b^T$ for some arc e_b, the probability $P_{\rho_0,a}(k)$ is simply the *ab*-entry of the following Schur product:

$$U^k \circ \overline{U^k};$$

we will refer to this matrix as the *mixing matrix* at step k.

What about the probability that the walker is on a vertex u? This is defined to be the sum of $P_{\rho_0,a}(k)$ over all outgoing arcs a of u. More generally, for any subset S of the arcs of X, the probability that the walker is on S at time k is

$$P_{\rho_0,S}(k) := \sum_{a \in S} P_{\rho_0,a}(k).$$

If ρ_S is the uniform mixed state over S, that is,

$$\rho_S := \frac{1}{|S|} \sum_{a \in S} e_a e_a^T,$$

then

$$P_{x_0,S}(k) = |S|\langle \rho_k, \rho_S \rangle. \tag{4.2.1}$$

This will be the main formula we use when dealing with the limiting distribution.

4.3 Average States and Average Probabilities

A well-known fact about classical random walks is that the probability distribution converges to a stationary distribution, under only mild conditions. Thus, it is natural to ask whether the state or the probability distribution converges in a quantum walk. Unfortunately, since U preserves the difference between states at two consecutive steps, neither ρ_k nor $P_{\rho_0,S}(k)$ converges, unless $\rho_1 = \rho_0$. (For a detailed explanation, see Aharonov et al [1].)

Nonetheless, the Cesàro sums of both $\{\rho_k\}$ and $\{P_{\rho_0,S}(k)\}$ exist. The first Cesàro sum,

$$\lim_{K \to \infty} \frac{1}{K} \sum_{k=0}^{K-1} \rho_k,$$

is called the *average state*; it was proposed by von Neumann as a first step towards thermalization [70]. We will give a formula for the average state using the spectral idempotents of U and apply it to find the second Cesàro sum, the *average probability*:

$$\lim_{K \to \infty} \frac{1}{K} \sum_{k=0}^{K-1} P_{\rho_0,S}(k).$$

4.3.1 Lemma. *Let U be a unitary matrix with spectral decomposition*

$$U = \sum_r e^{i\theta_r} F_r.$$

We have

$$\frac{1}{K}(U^k)\rho_0(U^k)^* = \sum_r F_r \rho_0 F_r + \frac{1}{K} \sum_{r \neq s} \left(\frac{1 - e^{iK(\theta_r - \theta_s)}}{1 - e^{i(\theta_r - \theta_s)}} \right) F_r \rho_0 F_s.$$

Proof By the spectral decomposition of U^k,

$$(U^k)\rho_0(U^k)^* = \sum_{r,s} e^{ik(\theta_r - \theta_s)} F_r \rho_0 F_s$$

$$= \sum_r F_r \rho_0 F_r + \sum_{r \neq s} e^{ik(\theta_r - \theta_s)} F_r \rho_0 F_s.$$

Hence

$$\frac{1}{K} \sum_{k=0}^{K-1} U^k \rho_0 (U^k)^* = \sum_r F_r \rho_0 F_r + \frac{1}{K} \sum_{r \neq s} \left(\sum_{k=0}^{K-1} e^{ik(\theta_r - \theta_s)} \right) F_r \rho_0 F_s$$

$$= \sum_r F_r \rho_0 F_r + \frac{1}{K} \sum_{r \neq s} \left(\frac{1 - e^{iK(\theta_r - \theta_s)}}{1 - e^{i(\theta_r - \theta_s)}} \right) F_r \rho_0 F_s. \qquad \square$$

4.3.2 Theorem. *Let U be a unitary matrix with spectral decomposition*

$$U = \sum_r e^{i\theta_r} F_r.$$

Given initial state ρ_0, the average state of the quantum walk with U as the transition matrix is

$$\lim_{K \to \infty} \frac{1}{K} \sum_{k=0}^{K-1} \rho_k = \sum_r F_r \rho_0 F_r.$$

Proof By Lemma 4.3.1, it suffices to prove that each entry in the residual

$$\frac{1}{K} \sum_{k=0}^{K-1} U^k \rho_0 (U^k)^* - \sum_r F_r \rho_0 F_r$$

is bounded by some constant independent of K. Indeed, for any K and any $r \neq s$,

$$\left| \frac{1 - e^{iK(\theta_r - \theta_s)}}{1 - e^{i(\theta_r - \theta_s)}} \right| \leq \frac{2}{|1 - e^{i(\theta_r - \theta_s)}|},$$

which only depends on r and s. $\qquad\square$

The map

$$\rho_0 \mapsto \sum_r F_r \rho_0 F_r$$

is known as the conditional expectation onto the commutant of U. We give another interpretation of this map from a channel viewpoint. (For background on quantum channels, see [11, 45, 55].) Since the eigenprojections satisfy

$$\sum_r F_r^* F_r = I,$$

the mapping on density matrices given by

$$\rho_0 \mapsto \sum_r F_r \rho_0 F_r^*$$

is a quantum channel. Therefore, the time-averaged state is effectively the image of the initial state passing through this channel.

The formula for the average probability now follows from Equation (4.2.1).

4.3.3 Theorem. *Let X be a digraph. Let U be a transition matrix of a quantum walk on X, with spectral decomposition*

$$U = \sum_r e^{i\theta_r} F_r.$$

Given initial state ρ_0 and a subset S of arcs, the average probability of the quantum walker being on S is

$$\lim_{K \to \infty} \frac{1}{K} \sum_{k=0}^{K-1} P_{\rho_0,S}(k) = |S| \sum_r \langle F_r \rho_0 F_r, \rho_S \rangle. \qquad \square$$

Two questions about the average probability are of interest: how fast does the partial sum

$$\frac{1}{K} \sum_{k=0}^{K-1} P_{\rho_0,S}(k)$$

converge, and when is the average probability distribution uniform? We will investigate these in the next two sections, respectively.

4.4 Mixing Times

Given $\epsilon > 0$, the *mixing time* $M_{\rho_0,S}(\epsilon)$ with respect to initial state ρ_0 and target arcs S is the smallest L such that for all $K > L$,

$$\left| \frac{1}{K} \sum_{k=0}^{K-1} P_{\rho_0,S}(k) - |S| \sum_r \langle F_r \rho_0 F_r, \rho_S \rangle \right| \leq \epsilon.$$

There are several variants of this definition. For instance, we may consider the mixing time conditioned on the initial state as being any standard basis vector:

$$M_S(\epsilon) := \sup\{M_{\rho_0,S}(\epsilon) : \rho_0 = e_a e_a^T \text{ for some arc } a\}.$$

For a more global purpose, we could look at the smallest L such that for all $K > L$, the average probability distribution over vertices is ϵ-close to the limiting distribution over vertices. In [1], Aharonov and colleagues studied the mixing time of the last type, and obtained an upper bound for a general graph. They further showed that the mixing time of a quantum walk on an n-cycle with the Hadamard coin is bounded above by $O(n \log n)$, giving a quadratic speedup over the classical random walk. We now extend some of their results on mixing times of the form $M_{\rho_0,S}(\epsilon)$.

4.4.1 Theorem. *Let X be a digraph. Let U be a transition matrix of a quantum walk on X, with spectral decomposition*

$$U = \sum_r e^{i\theta_r} F_r.$$

Given initial state ρ_0 and a subset S of arcs, the mixing time $M_{\rho_0,S}(\epsilon)$ satisfies

$$M_{\rho_0,S}(\epsilon) \leq \frac{2|S|}{\epsilon} \sum_{r \neq s} \frac{|\langle F_r \rho_0 F_s, \rho_S \rangle|}{|e^{i\theta_r} - e^{i\theta_s}|} \tag{4.4.1}$$

$$\leq \frac{2}{\epsilon} \sum_{r \neq s} \sum_{a \in S} \frac{\sqrt{(F_r)_{aa}(F_s)_{aa}}}{|e^{i\theta_r} - e^{i\theta_s}|} \tag{4.4.2}$$

$$\leq \frac{2|S|}{\epsilon} \sum_{r \neq s} \frac{1}{|e^{i\theta_r} - e^{i\theta_s}|} \tag{4.4.3}$$

$$\leq \frac{2\ell |S|}{\epsilon \Delta}, \tag{4.4.4}$$

where ℓ is the number of pairs of distinct eigenvalues, and

$$\Delta := \min\{|e^{i\theta_r} - e^{i\theta_s}| : r \neq s\}.$$

Proof From Lemma 4.3.1 we see that

$$\left|\frac{1}{K}P_{\rho_0,S}(k) - |S|\sum_r \langle F_r \rho_0 F_r, \rho_S \rangle\right| = \frac{|S|}{K}\left|\sum_{r \neq s} \frac{1 - e^{iK(\theta_r - \theta_s)}}{1 - e^{i(\theta_r - \theta_s)}} \langle F_r \rho_0 F_s, \rho_S \rangle\right|$$

$$\leq \frac{|S|}{K} \sum_{r \neq s} \left|\frac{1 - e^{iK(\theta_r - \theta_s)}}{1 - e^{i(\theta_r - \theta_s)}}\right| |\langle F_r \rho_0 F_s, \rho_S \rangle|$$

$$\leq \frac{2|S|}{K} \sum_{r \neq s} \frac{|\langle F_r \rho_0 F_s, \rho_S \rangle|}{|e^{i\theta_r} - e^{i\theta_s}|} \tag{4.4.5}$$

$$= \frac{2|S|}{K} \sum_{r \neq s} \frac{|\langle \rho_0, F_r \rho_S F_r \rangle|}{|e^{i\theta_r} - e^{i\theta_s}|}$$

$$\leq \frac{2|S|}{K} \sum_{r \neq s} \frac{\|F_r \rho_S F_s\|}{|e^{i\theta_r} - e^{i\theta_s}|}$$

$$\leq \frac{2}{K} \sum_{r \neq s} \sum_{a \in S} \frac{\sqrt{(F_r)_{aa}(F_s)_{aa}}}{|e^{i\theta_r} - e^{i\theta_s}|} \tag{4.4.6}$$

$$\leq \frac{2|S|}{K} \sum_{r \neq s} \frac{1}{|e^{i\theta_r} - e^{i\theta_s}|} \tag{4.4.7}$$

$$\leq \frac{2\ell |S|}{K \Delta}. \tag{4.4.8}$$

Thus, for all K such that

$$K > \frac{2|S|}{\epsilon} \sum_{r \neq s} \frac{|\langle F_r \rho_0 F_s, \rho_S \rangle|}{|e^{i\theta_r} - e^{i\theta_s}|},$$

the right-hand side of Inequality (4.4.5) is no more than ϵ. Similarly, the other three bounds follow from Inequalities (4.4.6), (4.4.7), and (4.4.8). $\qquad\square$

The last bound in Theorem 4.4.1 is equivalent to Lemma 4.3 in Aharonov and colleagues [1]. The other three bounds are stronger, but require more knowledge of the quantum walk besides the eigenvalues of U.

Below we present some data on the four upper bounds for two models on the circulant graph $X = X(\mathbb{Z}_n, \{1, -1, 2, -2\})$. Choose an initial state that concentrates on vertex 0, that is,

$$\rho_0 = \frac{1}{4} E_{00} \otimes J.$$

Let S_v denote the set of outgoing arcs of vertex v. For each upper bound $\beta_{\rho_0, S}$ in (4.4.1), (4.4.2), (4.4.3), and (4.4.4), we compute

$$\frac{\epsilon}{2} \sum_{v \in \mathbb{Z}_n} \beta_{\rho_0, S_v},$$

and store them in Table 4.1.

The models we consider are the arc-reversal Grover walk, which we introduced in Chapter 1, and the shunt-decomposition Grover walk, which we will introduce in Chapter 7. Let U_{ar} be the transition matrix of the arc-reversal Grover walk on X, and U_{sd} the transition matrix of the shunt-decomposition Grover walk on X. One can verify that if the spectral decomposition of U_{ar} is

$$U_{ar} = \sum_r \alpha_r F_r,$$

then the spectral decomposition of U_{sd} is

$$U_{sd} = \sum_r -\alpha_r F'_r,$$

where for any r, the eigenprojections F_r and F'_r have the same diagonal. Thus, the upper bounds (4.4.2), (4.4.3), and (4.4.4) are identical for both models. However, the last two columns in Table 4.1 indicate a difference between these two models – the shunt-decomposition Grover walk may have a lower mixing time than the arc-reversal Grover walk.

4.5 Average Mixing Matrix

Let X be a digraph. Let U be a transition matrix of a quantum walk on X, with spectral decomposition

$$U = \sum_r e^{i\theta_r} F_r.$$

Table 4.1 *Upper bounds for the mixing time on* $X(\mathbb{Z}_n, \{1, 2, -1, -2\})$

n	(4.4.4) ar/sd	(4.4.3) ar/sd	(4.4.2) ar/sd	(4.4.1) ar	(4.4.1) sd
6	1390.93	598.05	85.19	1.69	0.91
7	4620.05	1516.4	148.57	2.61	1.91
8	17771.88	3408.35	240.94	4.29	2.36
9	24838.95	3991.5	287.33	4.82	1.73
10	14285.23	3687.22	269.22	4.49	2.12
11	95452.33	9092.93	508.32	6.7	2.97
12	23505.04	4678.44	348.8	4.25	2.45
13	79048.14	13277.27	640.52	8.06	3.09
14	148284.47	19895.81	803.94	10.1	3.97
15	225507.28	16355.5	764.76	9.33	2.46
16	371901.16	34910.54	1211.24	13.57	4.96
17	2591443.27	65759.41	2127.89	24.26	5.54
18	330012.11	36141.49	1284.51	13.45	4.39
19	4854951.51	94822.86	2743.75	33.37	6.24
20	518641.81	51235.99	1562.68	15.33	5.37
21	848915.39	70921.92	1994.95	17.94	5.04
22	4443833.25	129338.06	3143.36	28.53	7.97
23	1651611.78	101994.04	2577.25	21.63	6.41
24	887647.03	76568.55	2185.39	18.05	6.69
25	1715366.87	103250.55	2603.42	20.88	5.9

In this section, we pay special attention to the average probability of the quantum walk from one arc a to another arc b, that is,

$$\sum_r \langle F_r e_a e_a^T F_r, e_b e_b^T \rangle.$$

Note that this is precisely the *ab*-entry of $\sum_r F_r \circ \overline{F_r}$. Following the approach for continuous quantum walks in [28], we let

$$\widehat{M} := \sum_r F_r \circ \overline{F_r}$$

and call $\widehat{M}$ the *average mixing matrix*. Theorem 4.3.3 implies that

$$\widehat{M} = \lim_{K \to \infty} \frac{1}{K} \sum_{k=0}^{K-1} U^k \circ \overline{U^k}.$$

In [28], several properties of the average mixing matrix for a continuous quantum walk are derived. Here we extend some of these results to discrete quantum walks.

The first observation is that $\widehat{M}$ is doubly stochastic. Moreover, since each F_r is Hermitian, $\widehat{M}$ is symmetric although $U^k \circ \overline{U^k}$ is not. Thus, we can view either the ath row or the ath column of $\widehat{M}$ as the average probability distribution given initial state $e_a e_a^T$.

For a continuous quantum walk, the average mixing matrix is proved to be positive semidefinite with eigenvalues no greater than one [28]. We show that the same statement holds for the discrete average mixing matrix.

4.5.1 Lemma. *The average mixing matrix $\widehat{M}$ of a quantum walk is positive semidefinite, and its eigenvalues lie in $[0, 1]$.*

Proof Since F_r is positive semidefinite, its complex conjugate $\overline{F_r}$ is positive semidefinite as well. Hence $F_r \otimes \overline{F_r}$ is positive semidefinite. As a principal submatrix of $F_r \otimes \overline{F_r}$, the Schur product $F_r \circ \overline{F_s}$ must also be positive semidefinite. Therefore, the eigenvalues of $\widehat{M}$ are non-negative. It follows from

$$ I = I \circ I = \left(\sum_r F_r \right) \circ \left(\sum_s \overline{F_s} \right) = \widehat{M} + \sum_{r \neq s} F_r \circ \overline{F_s} $$

and the positive-semidefiniteness of $F_r \circ \overline{F_s}$ that the eigenvalues of $\widehat{M}$ are at most 1. On the other hand, since $\widehat{M}$ is doubly stochastic, $\mathbf{1}$ is an eigenvector for $\widehat{M}$ with eigenvalue 1. $\qquad\qquad\square$

To measure the flatness of $\widehat{M} e_a$, we define its *entropy* to be the negative expectation of the logarithm of its entries, that is,

$$ -\sum_b \widehat{M}_{ab} \log(\widehat{M}_{ab}). $$

This quantity reaches maximum if and only if the probability distribution $\widehat{M} e_a$ is uniform. Likewise, the *total entropy* of $\widehat{M}$ is

$$ -\sum_{a,b} \widehat{M}_{ab} \log(\widehat{M}_{ab}); $$

it is maximized when the entire average mixing matrix is flat. In [9], Bai, Rossi, Cui, and Hancock proposed a graph signature based on the total entropy of continuous quantum walks. According to their experimental results, this entropic measure provides significant information on the properties of graphs.

A quantum walk with flat $\widehat{M}$ is said to admit *uniform average mixing*. According to the definition of $\widehat{M}$, uniform average mixing means that, in the limit, the walker has equal chance of being on any arc, no matter which arc she started with. In fact, as we will see later, something stronger is true when

$\widehat{M}$ is flat – the average probability distribution is uniform over all the arcs, regardless of the initial state.

While $\widehat{M}$ contains complete information on the average probabilities from arcs to arcs, one may be interested in average probabilities on the vertices as well. We say a quantum walk admits *uniform average vertex mixing* if the walker has equal chance of being on any vertex in the limit, regardless of the initial state.

Our next goal is to establish necessary and sufficient conditions for uniform average mixing to occur.

4.5.2 Lemma. *Let U be an $m \times m$ unitary matrix with spectral decomposition*

$$U = \sum_r e^{i\theta_r} F_r.$$

If ℓ_r is the multiplicity of the rth eigenvalue of U, then

$$\operatorname{tr}(\widehat{M}) \geq \frac{1}{m} \sum_r \ell_r^2.$$

Further, equality holds if and only if each idempotent F_r has constant diagonal.

Proof Since F_r is positive semidefinite, its diagonal entries are non-negative and

$$\operatorname{tr}(F_r) = \ell_r.$$

By Cauchy-Schwarz,

$$\operatorname{tr}(F_r \circ \overline{F_r}) \geq \frac{1}{m} \operatorname{tr}(F_r)^2 = \frac{1}{m} \ell_r^2.$$

Hence

$$\operatorname{tr}(\widehat{M}) \geq \frac{1}{m} \sum_r \ell_r^2.$$

Equality holds if and only if each F_r has constant diagonal ℓ_r/m. $\square$

4.5.3 Corollary. *Let U be a unitary matrix, and $\widehat{M}$ the associated average mixing matrix. Then $\operatorname{tr}(\widehat{M}) \geq 1$, with equality holding if and only if U has simple eigenvalues with flat eigenprojections.*

Proof Note that

$$\sum_r \ell_r = m.$$

By Lemma 4.5.2 and Cauchy-Schwarz,

$$\operatorname{tr}(\widehat{M}) \geq \frac{1}{m} \sum_r \ell_r^2 \geq 1.$$

Equality holds if and only if for all r, the idempotent F_r is a rank-one projection with constant diagonal, that is, each eigenvalue is simple with flat eigenprojections. $\qquad\square$

With all the tools established, we are ready to characterize uniform average mixing in discrete quantum walks.

4.5.4 Theorem. *Let U be a transition matrix of a quantum walk, and $\widehat{M}$ the associated average mixing matrix. The following statements are equivalent.*

(i) The quantum walk admits uniform average mixing.

(ii) $\operatorname{tr}(\widehat{M}) = 1$.

(iii) U has simple eigenvalues with flat eigenprojections.

Proof If uniform average mixing occurs, then all entries of $\widehat{M}$ are equal to $1/m$, so $\operatorname{tr}(\widehat{M}) = 1$. Hence (i) implies (ii). It follows from Corollary 4.5.3 that (ii) implies (iii). Now suppose (iii) holds. Then the spectral decomposition of U is

$$U = \sum_{r=1}^{m} e^{i\theta_r} F_r,$$

where for each r, all entries in F_r have the same absolute value. Hence

$$\widehat{M}_{ab} = \sum_{r=1}^{m} (F_r \circ \overline{F_r})_{ab} = \sum_{r=1}^{m} |(F_r)_{ab}|^2,$$

which does not depend on a and b. Therefore (iii) implies (i). $\qquad\square$

What about average probabilities on vertices, or subsets of arcs? The following result shows that if $\widehat{M}$ is flat, then the average probability that the walker is on some subset S of arcs depends only on the size $|S|$. In particular, uniform average mixing implies uniform average vertex mixing.

4.5.5 Theorem. *Let X be a digraph. If a quantum walk on X admits uniform average mixing, then for any initial state ρ_0 and any arc set S,*

$$\lim_{K \to \infty} \frac{1}{K} \sum_{k=0}^{K-1} P_{\rho_0, S}(k) = \frac{|S|}{nd}.$$

Proof Suppose uniform average mixing occurs. By Theorem 4.5.4, we can write the spectral decomposition of U as

$$U = \sum_{r=1}^{m} e^{i\theta_r} F_r,$$

where each F_r is a rank-one flat matrix. By Theorem 4.3.3,

$$\lim_{K \to \infty} \frac{1}{K} \sum_{k=0}^{K-1} P_{\rho_0, S}(k) = |S| \sum_{r=1}^{m} \langle F_r \rho_0 F_r, \rho_S \rangle$$

$$= |S| \sum_{r=1}^{m} \langle \rho_0, F_r \rho_S F_r \rangle$$

$$= \frac{|S|}{nd} \sum_{r=1}^{m} \langle \rho_0, F_r \rangle$$

$$= \frac{|S|}{nd} \langle \rho_0, I \rangle$$

$$= \frac{|S|}{nd}. \qquad \square$$

We note that simple eigenvalues and flat eigenvectors are sufficient, but not necessary for uniform average vertex mixing. Later in Chapter 7, we will construct an infinite family of quantum walks that admit uniform average vertex mixing, where the transition matrices do not have simple eigenvalues; these are quantum walks on circulant digraphs.

To end this section, we prove some algebraic properties of $\widehat{M}$. They rely on the well-known fact that a commutative semisimple matrix algebra with identity has a basis of orthogonal idempotents. In continuous quantum walks, similar results turn out to be quite useful in determining uniform mixing; see, for example, [34]. We hope the analogy in discrete quantum walks will be of use too.

4.5.6 Theorem. *Let U be the transition matrix of a quantum walk, and $\widehat{M}$ the associated average mixing matrix. If the entries of U are algebraic over $\mathbb{Q}$, then the entries of $\widehat{M}$ are algebraic over $\mathbb{Q}$.*

Proof Suppose U has algebraic entries. Then its eigenvalues are all algebraic. Let $\mathbb{F}$ be the smallest field containing the eigenvalues of U. Let $\mathcal{B}$ be the matrix algebra generated by U over $\mathbb{F}$. To show that $\mathcal{B}$ is semisimple, pick $N \in \mathcal{B}$ with

$N^2 = 0$. Since U is unitary, the algebra $\mathcal{B}$ is closed under conjugate transpose and contains the identity. It follows from $(N^*)^2 = 0$ that

$$
\begin{aligned}
0 &= \operatorname{tr}((N^*)^2 N^2) \\
&= \operatorname{tr}(N^* N N^* N) \\
&= \operatorname{tr}((N^* N)^* (N^* N)).
\end{aligned}
$$

Thus, $N^* N = 0$. Applying the trace again to $N^* N$, we see that $N = 0$. Therefore, the spectral idempotents F_r of U are polynomials in U with algebraic coefficients. Hence the entries in

$$
\widehat{M} = \sum_r F_r \circ \overline{F_r}
$$

are algebraic over $\mathbb{Q}$. $\qquad\square$

In continuous quantum walks, the entries of the average mixing matrix are all rational [28]. We show that the discrete average mixing matrix enjoys the same property, given that all entries of U are rational.

4.5.7 Theorem. *Let U be the transition matrix of a quantum walk, and $\widehat{M}$ the associated average mixing matrix. If the entries of U are rational, then the entries of $\widehat{M}$ are rational.*

Proof Let the spectral decomposition of U be

$$
U = \sum_r \alpha_r F_r.
$$

Let $\mathbb{F}$ be the smallest field containing the eigenvalues of U. Let σ be an automorphism of $\mathbb{F}$. Since U is rational, we have

$$
U = U^\sigma = \sum_r \alpha_r^\sigma F_r^\sigma.
$$

Moreover, since α_r^σ is also an eigenvalue of U, the set of idempotents $\{F_r\}$ is closed under field automorphisms. Thus,

$$
\widehat{M} = \sum_r F_r \circ F_r^T
$$

is fixed by all automorphisms of $\mathbb{F}$ and must be rational. $\qquad\square$

4.6 Continuous Quantum Walks

Alongside the discrete quantum walks we are studying, there is a second important class of quantum walks: *continuous quantum walks*. These are again based on graphs: if $A = A(X)$ then the *transition matrix* $U(t)$ for the continuous walk on X is given by

$$U(t) := \exp(itA), \qquad t \in \mathbb{R}.$$

Note that $U(t)$ is both symmetric and unitary. We define the *mixing matrix* $M(t)$ by

$$M(t) := U(t) \circ \overline{U(t)} = U(t) \circ U(-t).$$

If $U(t)$ describes the evolution of a quantum system with state space $\mathbb{C}^{V(X)}$, then $(M(t))_{i,j}$ is the probability that the outcome of a measurement in the standard basis is j, given that the initial state was i. There are two cases of particular interest to us:

(a) Some row of $M(t)$ is a standard basis vector.
(b) A row of $M(t)$ is constant (necessarily with all entries equal to $1/|V(X)|$).

In case (a) we have $M(t)_{a,b}$ for some pair of vertices a and b. If $a = b$, we say that the walk is *periodic at* a; if $a \neq b$, then we have *perfect state transfer* from a to b. In case (b) we say that we have *local uniform mixing* at the vertex a.

The simplest non-trivial example, K_2, reveals most of the properties of interest. In this case

$$A = \begin{pmatrix} 0 & 1 \\ 1 & 0 \end{pmatrix}$$

and $A^{2m} = I$ and $A^{2m+1} = A$. It follows that

$$U(t) = \begin{pmatrix} \cos(t) & i\sin(t) \\ i\sin(t) & \cos(t) \end{pmatrix}$$

and

$$M(t) = \begin{pmatrix} \cos^2(t) & \sin^2(t) \\ \sin^2(t) & \cos^2(t) \end{pmatrix}.$$

We see that K_2 is periodic at each of its vertices at time π and, at time $\pi/2$, we have perfect state transfer from vertex 1 to vertex 2, and vice versa. At time $\pi/4$ we have local uniform mixing from either vertex.

Although there are many similarities between continuous and discrete quantum walks, there are also significant differences. First there is a symmetry property: if, in a continuous walk, there is perfect state transfer from vertex a to vertex b at time t, then there will also be perfect state transfer from b to a at time t. For suppose

$$D_a = e_a e_a^T, \quad D_b = e_b e_b^T.$$

We have perfect state transfer from a to b at time t if

$$D_b = U(t)D_aU(-t)$$

and, if this holds, then

$$D_a = U(-t)D_bU(t).$$

Noting that D_a and D_b are real and taking the complex conjugates of both sides of this equation, we obtain

$$D_a = U(t)D_bU(-t).$$

This proves our claim; it also shows that if we have perfect state transfer as described, then

$$D_a = U(2t)D_aU(-2t),$$

whence X is periodic at a at time $2t$ (and similarly periodic at b). Compare this with discrete walk with transition matrix U. If

$$U^kD_aU^{-k} = D_b,$$

then

$$D_a = U^{-k}D_bU^k$$

but, since $\overline{U} = U^T$ (and U is not necessarily symmetric), we find only that

$$D_a = (U^T)^kD_b(U^T)^{-k} = (U^T)^kU^kD_aU^{-k}(U^T)^{-k}.$$

At this point we are stuck.

If $X \,\square\, Y$ denotes the Cartesian product of graphs X and Y, then

$$A(X \,\square\, Y) = A(X) \otimes I + I \otimes A(Y)$$

(which can serve as a definition of this product). It follows that

$$U_{X\square Y} = U_X(t) \otimes U_Y(t).$$

Using this we may show, for example, that the d-cube, the dth Cartesian power of K_2, admits perfect state transfer between vertices at distance d at time $\pi/2$, and uniform mixing at time $\pi/4$.

However, the transition matrix for an arc-reversal walk on $X \,\square\, Y$ has no useful expression in terms of the transition matrices of the factors.

Finally, we discuss the average mixing matrix of a continuous quantum walk, referring the reader to [28] for missing details. We define the *average mixing matrix* $\widehat{M}$ for a continuous quantum walk with mixing matrix $M(t)$ by

$$\widehat{M} = \lim_{T\to\infty} \frac{1}{T} \int_0^T M(t)\,dt.$$

Since

$$M(t) = U(t) \circ U(-t)$$
$$= \sum_{r,s} e^{it(\theta_r - \theta_s)} E_r \circ E_s$$
$$= \sum_r E_r^{\circ 2} + 2 \sum_{r<s} \cos((\theta_r - \theta_s)t) E_r \circ E_s,$$

it is not hard to show that

$$\widehat{M} = \sum_r E_r^{\circ 2}.$$

The spectral idempotents E_r are positive semidefinite and so, by a theorem of Schur, the matrices $E_r^{\circ 2}$ are positive semidefinite. It follows that $\widehat{M}$ is positive semidefinite. As the matrices $M(t)$ are doubly stochastic and as $\widehat{M}$ is a weighted average of the matrices $M(t)$, it follows that $\widehat{M}$ is doubly stochastic.

5

Covers and Embeddings

We introduce some of the theory of graph covers and graph embeddings.

5.1 Covers of Graphs

We will work with simple directed graphs, with no loops, and with each pair of vertices joined by at most one arc. If (u, v) and (v, u) are both arcs in X, we say that $\{u, v\}$ is an edge. Conversely, each edge gives rise to a pair of arcs. If X and Y are directed graphs, a map $\psi \colon V(X) \to V(Y)$ is a *homomorphism* if the image in Y of an arc of X is an arc in Y. We write $Y \to X$ to denote that there is a homomorphism from Y to X. The preimage $\psi^{-1}(u)$ in Y of a vertex u in X is the *fibre* of the cover at u. If there are no loops on u (as will always be the case), then the vertices in a fibre form a coclique in Y.

By way of example, there is a homomorphism $X \to K_m$ if and only if X is m-colourable.

A homomorphism ψ is a *local bijection* if, for each vertex x in X, the map from the out-neighbours of x to the out-neighbours of $\psi(x)$ is a bijection. We say that Y *covers* X if there is a surjective graph homomorphism ψ from Y to X that is locallly bijective. We refer to the pair (Y, ψ) as a *cover* of X. (There may be more than one covering map from Y to X.) Note that if ψ is surjective on vertices and locally bijective, it is also surjective on edges. When we say that two covers (Y_1, ψ_1) and (Y_2, ψ_2) are isomorphic, we mean that there is a graph isomorphism from Y_1 to Y_2 that maps each fibre of ψ_1 (bijectively) to a fibre of ψ_2.

5.1.1 Lemma. *Assume (Y, ψ) is a cover of X. If u and v are adjacent vertices in X, then the subgraph of Y induced by*

$$\psi^{-1}(u) \cup \psi^{-1}(v)$$

is a matching that pairs each vertex in $\psi^{-1}(u)$ with a vertex in $\psi^{-1}(v)$.

Proof Let $x \in \psi^{-1}(u)$. By the local bijection between neighbours of u and neighbours of x, there is exactly one neighbour y of x such that $\psi(y) = v$. Similarly, every vertex in $\psi^{-1}(v)$ is adjacent to exactly one vertex in $\psi^{-1}(u)$. $\qquad\square$

5.1.2 Corollary. *If X is connected and (Y, ψ) is a cover of X, all fibres of the cover have the same size. Each connected component of Y is a cover of X; its covering map is the restriction of ψ to the component.*

Proof If X is connected, then for each pair of vertices u and v, there is a path from u to v. By the preceding lemma, $\psi^{-1}(u)$ and $\psi^{-1}(v)$ have the same size. Let Z be a connected component of Y. Note that for each vertex z of Z, the map ψ restricted to Z is a local bijection between the neighbours of z and the neighbours of $\psi(z)$. $\qquad\square$

We define the *covering group* of a cover Y to be the set of automorphisms of Y that map each fibre to itself.

5.1.3 Lemma. *Let Y be a connected cover of X. An automorphism of Y that maps each fibre to itself and fixes a vertex must be the identity.*

Proof Let γ be an automorphism of Y that maps each fibre to itself. Suppose γ fixes the vertex u. Then γ permutes the neighbours of u. On the other hand, in each fibre, there is at most one neighbours v of u, so $h(v) = u$. Hence, h acts as the identity on the component of Y containing u. Since Y is connected, h must be the identity. $\qquad\square$

This lemma implies that the covering group acts fixed-point freely, whence all its orbits have the same length and this length is the order of the covering group. Consequently, the order of the covering group divides the index of the cover. A cover is said to be *regular* if the order of its covering group is equal to the index of the cover.

Any cover of index two is regular.

5.2 Constructing Covers

By virtue of Lemma 5.1.1, we may specify a cover of X by an *arc function*. We define this to be a map ψ from the arcs of X in some group such that if ab is an arc in X, we have

$$\psi(a,b)\psi(b,a) = 1.$$

Since a matching between two copies of $\{1,\dots,r\}$ defines a permutation, we see that each cover of index r of a graph X corresponds to an arc function on X taking values in $\mathrm{Sym}(r)$.

This provides us with a simple way to construct all possible covers of a graph, which we describe now.

Let X be a graph and let ψ be an arc function on X taking values in the symmetric group $\mathrm{Sym}(r)$. Let X^ψ denote the graph with vertex set

$$V(X) \times \{1,\dots,r\},$$

where there is an edge from (u,i) to (v,j) if (u,v) is an arc of X and $\psi(u,v)$ maps i to j.

The reader might verify that the map that sends a vertex (u,i) in X^ψ to u is a locally bijective graph homomorphism, and therefore X^ψ covers X.

As an example, consider the complete graph K_4, as shown in Figure 5.1. We can let ψ be the constant arc function that sends every arc to $(1,2) \in \mathrm{Sym}(2)$. Then the double cover K_4^ψ is isomorphic to the cube, as shown in Figure 5.2.

If f maps each arc of X to the identity, then X^f consists of r vertex-disjoint copies of X.

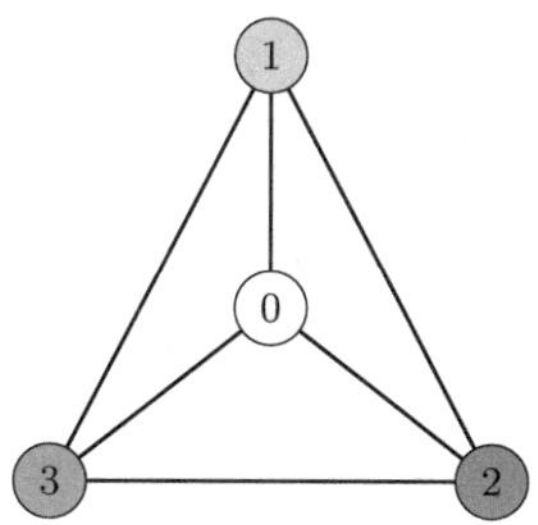

Figure 5.1 K_4

                    Covers and Embeddings

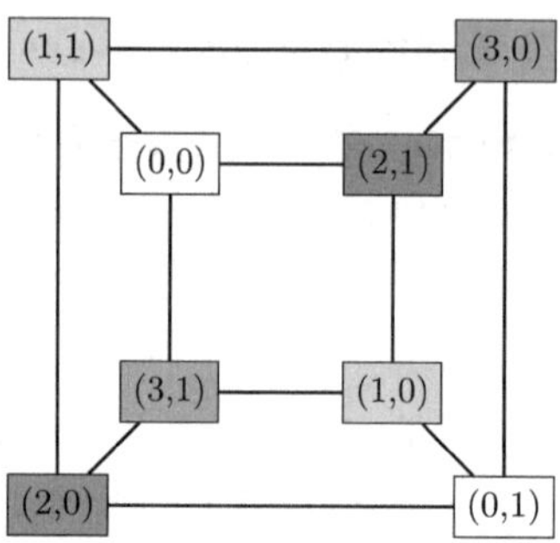

Figure 5.2 A double cover of K_4

5.3 Equivalence of Arc Functions

If f is an arc function of X with index r and

$$w = (w_0, \ldots, w_m)$$

is a walk on X, we define

$$f(w) = f(w_0, w_1) \cdots f(w_{m-1}, w_m).$$

Hence we have extended f from a function on arcs (walks of length one) to a function on walks.

Assume f is an arc function of index r on X, and for each vertex u of X, let σ_u be an element of $\mathrm{Sym}(r)$. Let g be the function on the arcs of X whose value on the arc (v, w) is given by

$$g(v, w) = \sigma_w f(v, w) \sigma_v^{-1}.$$

Then g is an arc function in our usual sense. We call the function that assigns the permutation σ_u to the vertex u a *switching*. The set of all switchings forms a group, isomorphic to the direct product of $|V(X)|$ copies of $\mathrm{Sym}(r)$. Two arc functions are *switching equivalent* if we can derive one from the other by applying a sequence of switchings.

5.3.1 Lemma. *If f and g are equivalent arc functions, they take the same values on each closed walk.* ☐

5.3.2 Theorem. *If f and g are switching equivalent arc functions on X, then X^f and X^g are isomorphic as covers.*

Proof Since f and g are switching equivalent, for each u there is a switching σ_u such that, for each arc (v, w),

$$g(v, w) = \sigma_w f(v, w) \sigma_v^{-1}.$$

Define $\gamma : V(X^f) \to V(X^g)$ by $\gamma((u, i)) = (u, \sigma_u(i))$. Then for each arc (v, w) of Y and each i,

$$g(u, v)\sigma_v(i) = \sigma_w f(v, w)(i).$$

Hence f is a homomorphism that maps each fibre of X^f bijectively to each fibre of X^g. $\qquad\square$

Now assume X is connected and choose a vertex u of X and spanning tree T. Let f be an arc function on X. If $v \in V(X)$, there is a unique path in T from u to v; let σ_v be the value of f on the associated walk and set σ_u equal to 1. This defines a switching determined by the pair (T, u).

5.3.3 Lemma. *Let X be connected, and let u be a vertex and T be a spanning tree in X. Let f be an arc function on X. Then the image of f under the switching associated with (T, u) is an arc function that takes the value 1 on each arc of T.* $\qquad\square$

We say that the image of f under (T, u) switching is *normalized* relative to T. We note that the normalization does not depend on the choice of root u.

5.3.4 Lemma. *If T is a tree, any connected cover of T is isomorphic to T.*

5.4 Reduced Walks

We work with strongly connected directed graphs. Note that, viewed as a directed graph, a graph is connected if and only if it is strongly connected. A closed walk in D is *reduced* if it does not contain a subsequence of the form xyx. If a closed walk does contain such a subsequence, we can shorten the walk by replacing xyx with x. We call this an *elementary reduction*, and a *reduction* is a sequence of elementary reductions.

5.4.1 Lemma. *If X is a directed graph, each walk has a unique reduction.*

Proof Let z be a walk in X, and let x and y be walks each obtained from x by an elementary reduction. Then it is an easy exercise to verify that there is a walk w that can be obtain from both x and y by elementary reductions.

Now we proceed by induction on the length of z. Suppose z reduces to the walks w_1 and w_2, and let x_1 and y_1 respectively be the walks produced in the step of these two reductions. By the first paragraph x_1 and y_1 have a common elementary reduction, which we denote by u, and, induction, u is the unique reduced walk derived from x_1 and from x_2. It follows that u is the unique reduced walk derived from z. $\qquad\square$

Say that two walks are *elementarily equivalent* if one can be obtained from the other by an elementary reduction; this is an equivalence relation on walks, and we refer to its transitive closure as *equivalence*.

5.4.2 Corollary. *Each equivalence class of closed walks in a directed graph X contains a unique reduced walk.* $\qquad\square$

We define a directed graph on the reduced walks in X that start at a given vertex v by defining a pair of reduced walks (x, y) to be an arc if $y = xa$ for some vertex a. Denote this graph by $T(X, v)$. Note that if reduced walks α and β are adjacent in $T(X, v)$, then the lengths of α and β differ by 1, from which it follows that $T(X, v)$ is bipartite.

5.4.3 Lemma. *If X is a directed graph and $v \in V(X)$, then $T(X, v)$ is a tree.*

Proof If α is a reduced walk of length k starting at v and $k \geq 1$, there is a unique reduced walk β of length $k - 1$ adjacent to α in $T(X, v)$.

Let $\tau(w)$ denote the last vertex in the walk w. If

$$w_0, \ldots, w_m$$

is a sequence of reduced walks starting at v such that $(w_i, w_i + 1)$ is an arc in $T(X, v)$ for each i, then the sequence of vertices

$$\tau(w_0), \ldots, \tau(w_m)$$

determines the sequence of walks, and the reduction of this sequence of vertices is a reduced walk that is equal to w_m.

We prove by contradiction that $T(X, v)$ has no directed cycles of length greater than two. Suppose

$$C = (\alpha_0, \alpha_1, \ldots, \alpha_0)$$

is a cycle in $T(X, v)$, and choose i so that the length ℓ of α_i is maximal. Then the walks α_{i-1} and α_{i+1} must have length $\ell - 1$, whence we find that

$\tau(\alpha_{i-1}) = \tau(\alpha_{i+1})$. Consequently $\alpha_{i-1} = \alpha_{i+1}$, contradicting the fact that C is a cycle. $\qquad\square$

We extend the definition of covers to directed graphs: a map f from $V(Y)$ to $V(X)$ is a homomorphism if the image of an arc is an arc. We define a homomorphism from Y to X to be a local isomorphism if, for each vertex y in Y, it is a bijection from the out-neighbours of y to the out-neighbours of $f(y)$.

5.4.4 Lemma. *The map that takes a reduced walk starting at v to its last vertex is a covering map from $T(X,v)$ to X.* $\qquad\square$

5.5 Products and Universal Covers

We will use the direct product of graphs to define a product of covers. We recall that the direct product $Y \times Z$ has vertex set $V(Y) \times V(Z)$, and vertices (y_1, z_1) and (y_2, z_2) in $Y \times Z$ are adjacent if and only if $y_1 y_2$ is an edge in Y and $z_1 z_2$ is an edge of Z. So

$$A(Y \times Z) = A(Y) \otimes A(Z).$$

There are two coordinate projections, mapping (y, z) to y and z; We denote them respectively by ρ_1 and ρ_2. The *diagonal* of the direct square $X \times X$ is the subgraph induced by the vertices (x, x) for x in $V(X)$; it is isomorphic to X.

5.5.1 Lemma. *The coordinate projections on the direct product are graph homomorphisms.* $\qquad\square$

Now suppose (Y, f) and (Z, g) are covers of X. Then the map from $Y \times Z$ to $X \times X$ that sends (y, z) in $V(Y \times Z)$ to $(f(y), g(z))$ in $X \times X$ is a graph homomorphism. We define the product of the two covers to be the preimage in $Y \times Z$ of the diagonal of $X \times X$, that is, it is the subgraph of $Y \times Z$ induced by the vertices in the set

$$P = \{(y, z) : f(y) = g(z)\}.$$

We see that $\rho_1(P) = V(Y)$ and $\rho_2(P) = V(Z)$. We will call the subgraph of $Y \times Z$ induced by P the *product* of the two covers.

5.5.2 Lemma. *Assume (Y, f) and (Z, g) are covers of X, and let P be the preimage of the diagonal of $X \times X$ under the map from $Y \times Z$ to $X \times X$. Then the restrictions to P of the projections π_1 and π_2 are covering maps to Y and Z, and* *the composition of these restrictions with f and g are equal and are covering maps of X.* $\square$

We have seen that if X is connected and $u \in V(X)$, the tree $T(X, u)$ on reduced walks in X starting at u covers X.

5.5.3 Theorem. *Assume X is connected and $u \in V(X)$. If (Y, f) covers X, then $T(X, u)$ covers Y.*

Proof Let Z denote the product of the two covers (Y, f) and $T(u, v)$. Then Z covers Y and $T(u, v)$ and, since $T(X, u)$ is a tree, each component of Z is isomorphic to $T(X, u)$. Therefore, $T(X, u)$ covers Y. $\square$

The composition of the covering map from $T(X, u)$ to Y with f is the canonical map from $T(X, u)$ to X; we leave this as an exercize. It is because of this theorem that $T(X, u)$ is called the universal cover of X.

The theory of covers we have outlined above can be extended to directed graphs. For convenenience we assume that directed graphs are simple – no loops and at most one arc between any pair of vertices. Such a graph has an associated undirected graph, with the same vertex set, and with two vertices adjacent in the graph if and only if they are joined by an arc in the directed graph. This graph is often referred to as the *underlying graph* (although strictly speaking it is overlying). Assume D is a simple directed graph with associated graph X. If ψ is an arc function on D, it is straightforward to extend ψ to an arc function on X, and then the cover D^ψ of D is a simple directed graph with associated graph X^ψ.

5.6 Graph Embeddings

A *cellular embedding* of a graph X in a surface is a drawing of a graph on the surface, such that edges only meet at vertices and, if we delete from the surface all points on vertices and edges of X, each connected component is topologically a disc. We will use *embedding* to denote cellular embedding. A drawing of a disconnected graph in the plane is **not** cellular, even if no edges cross. The regions into which the surface is divided by the vertex and edges of an embedded graph are the *faces* of the embedding. The boundary of each face determines two (oppositely directed) closed walks in X; we call these *boundary walks*. We would prefer that these boundary walks be cycles, which they often are. However any drawing of a tree in the plane without crossings is cellular, with one face, and the boundary walk is not a cycle. If each face is bounded by a cycle, we say that the embedding is *circular*.

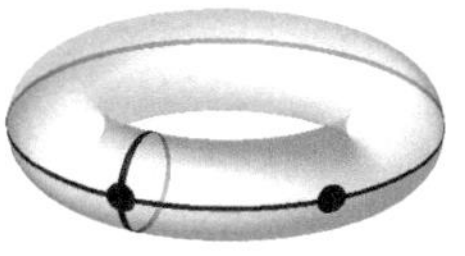

(a) A cellularly embedded graph *G*.

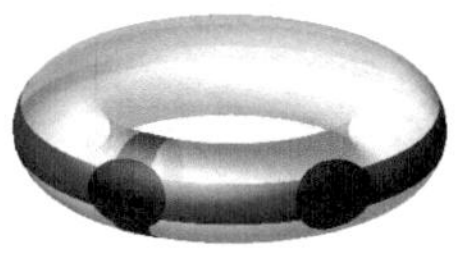

(b) *G* as a band decomposition.

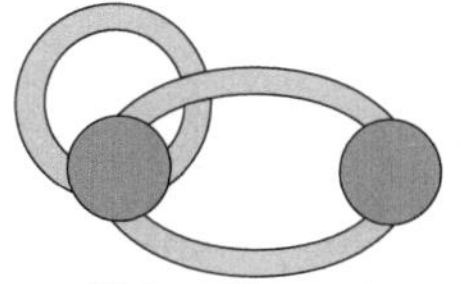

(c) *G* as a ribbon graph.

Figure 5.3 Relation between embeddings and Ribbon graphs; figures taken from Ellis-Monaghan and Moffatt [22]

5.6.1 Ribbon Graphs and Gems

A *ribbon graph* is a structure we derive from an embedding. Suppose we have an embedding of X in some surface. We enlarge each vertex into a small disc, and each edge into a rectangular strip. The relation between embeddings and ribbon graphs is illustrated in Figure 5.3.

If the embedding is circular, we may construct from its ribbon graph an edge 3-coloured cubic graph, called a *graph-encoded map*, or *gem*. The vertices of the new graph are the corners of the rectangles corresponding to the edges, and the four sides of a rectangle provide four edges, two long and two short. The disc corresponding to a vertex of degree k determines a cycle of length $2k$, where every second edge of the cycle is the short edge of a vertex rectangle; we refer (for now) to the other edges on the disc as open. If we label the long edges with 0, the short edges with 2, and the open edges with 1, we arrive at an edge three-coloured cubic graph, with edge labels coming from $\{0, 1, 2\}$.

Note that our gem determines the ribbon graph we derived it from and that this graph, in turn, determines the embedding of X in the surface.

An edge-colouring of a cubic graph consists of three perfect matchings, which we may view as permutations of the vertices. If our graphs arise from an embedding, its vertices are the corners of the ribbon graph. We can define the corners in another way, as follows. A *flag* of a map is a triple,

$$(\text{vertex}, \text{edge}, \text{face}),$$

where the vertex is contained in the edge, and the edge in the face. Clearly the topological concept of a corner corresponds to the combinatorial concept of a flag.

Figure 5.4 gives the planar embedding of C_3, where each dot represents a flag.

For each flag (u, e, f), let u' be the other endpoint of e, let e' be the other edge in f that is incident to u, and let f' be the other face that contains e. Define three functions

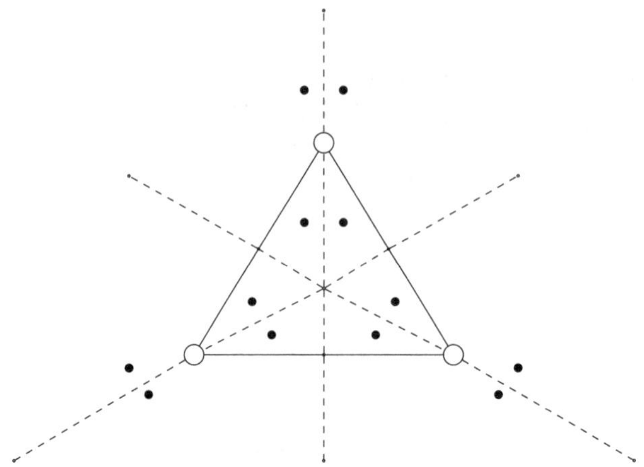

Figure 5.4 Planar embedding of C_3 and the flags

$$\tau_0\colon (u,e,f) \mapsto (u',e,f),$$
$$\tau_1\colon (u,e,f) \mapsto (u,e',f),$$
$$\tau_2\colon (u,e,f) \mapsto (u,e,f').$$

We have the following observations.

(i) τ_0, τ_1, τ_2 are fixed-point-free involutions.

(ii) $\tau_0\tau_2 = \tau_2\tau_0$, and $\tau_0\tau_2$ is fixed-point-free.

(iii) The group $\langle \tau_0, \tau_1, \tau_2 \rangle$ acts transitively on the flags.

If we join two flags in Figure 5.4 by an edge whenever they are swapped by one of τ_0, τ_1, and τ_2, then we obtain a cubic graph with a 3-edge colouring, as shown in Figure 5.5.

In general, for a circular embedding $\mathcal{M}$, a gem is a cubic graph with a 3-edge colouring, where the vertices are the flags, and the 3-edge colouring is induced by the three involutions τ_0, τ_1, and τ_2, as described above. The concept of gem was first introduced by Lins. In [52], he proved the following characterization of orientability.

5.6.1 Theorem. *An embedding is orientable if and only if the gem is bipartite.* $\qquad\square$

An automorphism of an embedding $\mathcal{M}$ is a permutation of the vertices that preserves the vertex-edge-face incidences. We will denote the automorphism

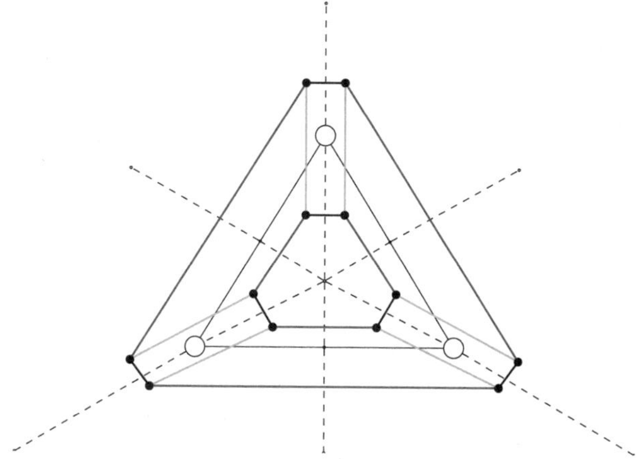

Figure 5.5 Planar embedding of C_3 and the gem

group of $\mathcal{M}$ by Aut($\mathcal{M}$). Note that Aut($\mathcal{M}$) is isomorphic to a subgroup of the automorphism group of the gem.

5.6.2 Lemma. *Let $\mathcal{M}$ be a circular embedding of a graph X, and let Y be the associated gem. For any flag (u, e, f), the stabilizer $\mathrm{Aut}(\mathcal{M})_{(u,e,f)}$ acts semiregularly on the flags.*

Proof Let the neighbours of (u, e, f) be (u', e, f), (u, e', f), and (u, e, f'). Let g be an automorphism of $\mathcal{M}$ that fixes (u, e, f). As g preserves the vertex-edge incidences, it must send u' to an endpoint of e that is not u. Hence, $(u')^g = u$. Similarly, as g preserves the edge-face incidences, it must send f' to itself. This leaves the only choice for $(u, e', f)^g$, that is, (u, e', f). $\qquad\square$

5.6.2 Rotation Systems

We have seen how to describe graph embeddings using ribbon graphs or gems. There is an alternative approach which provides a more compact representation. (A disadvantage is that it is less convenient if we are dealing with embeddings in non-orientable surfaces.)

A *rotation system* on a set F is a pair (σ, τ) of permutations of F such that

(a) σ is a fixed-point free involution.

(b) $\langle \sigma, \tau \rangle$ acts transitively on F.

From (a) we see that $|F|$ must be even. We can construct a rotation system from an embedding of a graph X in an orientable surface. Define F to be the set of arcs of X and let σ be the permutation of F that maps each arc (u, v) to (v, u). Define τ to be the permutation that maps an arc (u, v) to the arc (u, w) that follows (u, v) in the clockwise ordering of the arcs on u. Then (σ, τ) is a rotation system.

If (σ, τ) is a rotation system, we may view the cycles of τ as vertices, and declare two cycles to be adjacent if one contains an arc (u, v) and the other the arc (v, u).

For any arc (u_1, u_2), consider the walk

$$(u_1, u_2), (u_2, u_3), (u_3, u_4), \ldots, (u_{k-1}, u_k), \ldots$$

where two consecutive arcs are related by

$$(u_j, u_{j+1}) = \tau\sigma(u_{j-1}, u_j).$$

Since the graph is finite, eventually this walk will meet an arc that is already taken. Moreover, the first arc that is used twice must be (u_1, u_2), as $\tau\sigma$ is a permutation. Therefore this walk is closed with no repeated arc. All closed walks arising in this way partition the arcs of X; they are called the *facial walks*. For each facial walk of length k, we associate it with a polygon with k sides, labeled by the arcs in the same order as they appear in the walk. We then "glue" each two sides of these polygons labeled by the same edge. This results in an embedding of the graph onto an orientable surface.

5.6.3 Voltage Graphs

A *voltage graph* of X is an r-fold cover $Y = X^\phi$, where the image of the arc function ϕ is a subgroup $\Gamma \leq \mathrm{Sym}(r)$ of order r, and

$$V(Y) = V(X) \times \Gamma, \quad E(Y) = E(X) \times \Gamma.$$

Voltage graphs correspond to normal covers [40], and have been extensively studied. We only state one property that voltage graphs satisfy; for more background, see Gross and Tucker [38].

5.6.3 Theorem. *Let X be a graph. Let Z be a k-cycle in X. Let $Y = X^\phi$ be a voltage graph of order r. If $\phi(Z)$ has order ℓ, then the components of $F(Z)$ consist of r/ℓ cycles, each of length $k\ell$.* $\qquad\square$

Given an orientable embedding $\mathcal{M}_X$ of X, and a covering map ψ from a connected graph Y to X, we define an orientable embedding $\mathcal{M}_Y$ of Y by specifying its facial walks; such an embedding will be called the *embedding induced by* $(\mathcal{M}_X, \psi)$, or the *embedding induced by* $(\mathcal{M}_X, \phi)$ if ϕ is the corresponding arc function. Let W be a facial walk of $\mathcal{M}_X$ starting at vertex u. Clearly, the preimage $\psi^{-1}(W)$ consists of walks starting and ending in the fiber $\psi^{-1}(u)$, and each arc of Y appears in at most one of these walks. Then, the facial walks of $\mathcal{M}_Y$ are exactly the closed walks in the preimages of the facial walks of $\mathcal{M}_X$. In the previous example, the planar embedding of K_4 gives rise to an embedding of the cube on the torus, with four faces, each of length 6.

5.7 Self-Dual Embeddings

An embedding $\mathcal{M}$ of a graph X is *graph self-dual* if the dual graph X^* is isomorphic to X. For the complete graph K_n, the dual graph is regular on n vertices if and only if $\mathcal{M}$ is graph self-dual. If this embedding is circular, in addition to being graph self-dual, then the vertex-face incidence structure is the complement of a trivial design, that is, C can be obtained from $J - I$ by permuting the rows and columns.

Using Euler's formula, one can show that K_n has a graph self-dual embedding only if $n \equiv 0, 1 \pmod 4$. The other direction requires clever constructions, and has been proved several times independently. For one of these treatments, see White [72].

5.7.1 Theorem. *The complete graph K_n has a graph self-dual orientable embedding if and only if $n \equiv 0, 1 \pmod 4$.* $\qquad\square$

The smallest two examples are the planar embedding of K_4 and the toroidal embedding of K_5 as shown in Figure 5.6.

However, not all graph self-dual embeddings of K_n are circular. In fact, such constructions are only known for K_n with n a prime power, and are due to Biggs [12]. We describe his rotation systems in the following theorem.

5.7.2 Theorem. *Let $n = p^k$ for some prime p. Let g be a primitive generator of the finite field $\mathbb{F}$ of order n. For each element u in $\mathbb{F}$, define the cyclic permutation*

$$\pi_u := (v + g^0, v + g^1, \ldots, v + g^{n-2}).$$

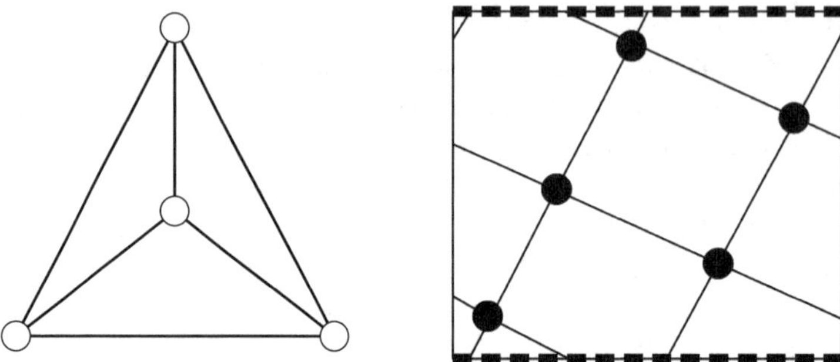

Figure 5.6 Self-dual embeddings of K_4 and K_5

Let τ be the permutation on the arcs that sends arc (u, v) to arc $(u, \pi_u(v))$. Let σ be the permutation that reverses each arc. Then the rotation system (σ, τ) gives a circular embedding of K_n.

Proof The complete graph K_n can be viewed as a Cayley graph over $\mathbb{F}$. Clearly, π_u is a permutation on the neighbours of u. Further, the facial walk containing arc $(v, v + g^0)$ visits vertices in the following order:

$$v, v + g^0, v + g^0 - g^1, v + g^0 - g^1 + g^2, \ldots.$$

Since

$$\sum_{j=0}^{m-1} (-g)^j = \frac{1 - (-g)^m}{1 + g}$$

is distinct for $m = 0, 1, \ldots, n - 2$, this facial walk has length $n - 1$ with no vertex repeated. Therefore the embedding is circular and graph self-dual. $\square$

5.8 Cayley Maps

A *Cayley map* for a group Γ is an embedding whose automorphism group contains a subgroup isomorphic to Γ acting regularly on the vertices. In this case, the underlying graph is a Cayley graph over Γ, and for each element g in the connection set, if τ sends the arc $(1, g)$ to $(1, h)$, then it sends the arc (u, gu) to (u, hu). In other words, a Cayley map is determined by a cyclic permutation on the connection set.

The following are two characterizations for a finite group Γ to have a regular Cayley map. We refer the readers to Richter and colleagues [60] and Conder and colleagues [17] for more details.

5.8.1 Theorem (Conder et al.). *The finite group Γ has a regular Cayley map if and only if there is a finite group $G = \langle \rho, \lambda \rangle$, with $\lambda^2 = 1$, such that G has a complementary factorization as $\Gamma \langle \rho \rangle$ with $A \cap \langle \rho \rangle = \{1\}$. The underlying group has multiple edges if and only if $\langle \rho \rangle$ contains a non-trivial normal subgroup of G.* $\qquad\square$

5.8.2 Theorem (Conder et al.). *The finite group Γ has a regular Cayley map if and only if there is a skew morphism f for Γ and an element g of Γ, such that the orbit of g under f generates Γ and contains g^{-1}.* $\qquad\square$

5.8.3 Corollary. *Let X be a Cayley graph over an abelian group Γ. Let $U = R(C \otimes I)$ be the transition matrix. Then the eigenvalues of U are precisely the eigenvalues of $R_\chi C$, as χ ranges over all linear characters of Γ.* $\qquad\square$

5.9 Regular Embeddings

Any automorphism of a circular embedding $\mathcal{M}$ is determined by its effect on one flag. Hence,

$$|\mathrm{Aut}(M)| \leq 4|E|.$$

When the equality holds, we say $\mathcal{M}$ is *regular*.

If the underlying surface is orientable, we may also consider the subgroup of automorphisms that preserve the orientation, denoted $\mathrm{Aut}^o \mathcal{M}$. Assign a consistent orientation to the faces of $\mathcal{M}$. Given any edge $e = \{u, v\}$ and an endpoint u, there is exactly one face f that uses the arc (u, v), and so

$$|\mathrm{Aut}^o \mathcal{M}| \leq 2|E|.$$

When the equality holds, we say $\mathcal{M}$ is *orientably regular*, or *rotary*.

5.9.1 Lemma. *Let M be an embedding defined by the rotation system (σ, τ). Then $\mathrm{Aut}^o \mathcal{M}$ is isomorphic to the centralizer of $\langle \sigma, \tau \rangle$.*

Proof Let P be an orientation-preserving automorphism. Then P induces a permutation θ of the arcs by

$$\theta(u, v) = (P(u), P(v)).$$

We show that θ commutes with σ and τ. First, for any arc (u, v),

$$\sigma\theta(u, v) = \tau(P(u), P(v)) = (P(v), P(u)) = \theta(v, u) = \theta\sigma(u, v).$$

This shows that $\tau\sigma = \sigma\tau$. Now suppose $\tau(u, v) = (u, w)$. Since P preserves the orientation,

$$\tau\theta(u, v) = \tau(P(u), P(v)) = (P(u), P(w)) = \theta(u, w) = \theta\tau(u, v),$$

and so $\tau\theta = \theta\tau$.

Conversely, let θ be a permutation of arcs that commutes with σ and τ. Fix a vertex u, and let v and w be two of its neighbours. Then some power of τ, say τ^j, maps (u, v) to (u, w). As

$$\tau^j\theta(u, v) = \theta\tau^j(u, v) = \theta(u, w), \tag{5.9.1}$$

the restriction of θ to the tails of the arcs defines a permutation of the vertices. Let P denote such permutation; we show that P is an orientation-preserving automorphism of $\mathcal{M}$. Indeed, if u is adjacent to v, and

$$\theta(u, v) = (P(u), a), \quad \theta(v, u) = (P(v), b),$$

then

$$(a, P(u)) = \sigma\theta(u, v) = \theta\sigma(u, v) = \theta(v, u) = (P(v), b),$$

from which we see that $P(u)$ is adjacent to $P(v)$. Moreover, the restriction of θ to the tails coincides with the restriction of θ to the heads, and so by Equation (5.9.1),

$$\tau^j(P(u), P(v)) = (P(u), P(v)).$$

This shows that P preserves the orientation. $\qquad\square$

5.9.2 Corollary. $\mathcal{M} = (\sigma, \tau)$ *is orientably regular if and only if the group* $\langle\sigma, \tau\rangle$ *acts regularly on the arcs.*

Biggs showed that K_n has a regular embedding if and only if n is a prime power, and every regular embedding of K_n must arise from the rotation system described in Theorem 5.7.2.

5.10 Quantum Rotation Systems

Consider a rotation system (σ, τ). Each orbit of the group $\langle\sigma\tau\rangle$ consists of the arcs used by some face, and so by Burnside's Lemma, the number of faces equals the average number of fixed points of $\langle\sigma\tau\rangle$. We express this idea in matrix form. Let R and S be the permutation matrices associated with σ and τ, respectively. Let $U = RS$, and assume U has order ℓ. Then

$$\frac{1}{\ell}\sum_{t=0}^{\ell-1}\mathrm{tr}(U^t)$$

is the number of faces. Notice that R coincides with the arc-reversal matrix, S serves as a permutation coin matrix, and

$$\frac{1}{\ell}\sum_{t=0}^{\ell-1}\mathrm{tr}(U^t) = \lim_{T\to\infty}\frac{1}{T}\sum_{t=0}^{T-1}\mathrm{tr}(U^t \circ U^t) = \mathrm{tr}(\widehat{M}).$$

Hence, a rotation system defines an arc-reversal walk (although somewhat uninteresting), and its average mixing matrix contains useful information about the embedding.

Motivated by this, we introduce a quantum analogue of the rotation system. Given the same R and S, we say (R, C) is a *quantum rotation system* if C is unitary and $C = p(S)$ for some polynomial $p(\cdot)$. If the graph is k-regular, and

$$C = \sum_{j:\gcd(j,k)=1} c_j S^j$$

for some coefficients c_j, then U lies in the span of

$$\{RS^j : \gcd(g,j) = 1\},$$

which represents a set of rotation systems of the graph. Thus, we may think of any quantum rotation system as a "superposition" of rotation systems. The trace of the average mixing matrix, $\mathrm{tr}(\widehat{M})$, will be the quantum analogue of the number of faces.

Lemma 4.5.2 gives a lower bound for $\mathrm{tr}(\widehat{M})$ and when the bound is attained. We apply this to quantum rotation systems whose underlying embeddings are orientably regular.

5.10.1 Lemma. *Let $\mathcal{M}$ be an orientably regular embedding. Let $U = RC$ for some quantum rotation system (R, C). Then*

$$\mathrm{tr}(\widehat{M}) = \frac{1}{m}\sum_r \ell_r^2,$$

where m is the number of arcs, and ℓ_r is the multiplicity of the rth eigenvalue of U. In particular, $\mathrm{tr}(\widehat{M})$ is rational.

Proof The algebra generated by $\langle R, S \rangle$ over $\mathbb{C}$ is a homogeneous coherent algebra, so each spectral idempotent of U has constant diagonal. The result follows from Lemma 4.5.2. $\square$

6

Vertex-Face Walks

6.1 Introduction

In this chapter, we construct a new discrete quantum walk from an orientable embedding of a graph. Roughly speaking, the walk is defined by two partitions of the arcs: one based on the faces, and one on the vertices. To illustrate the idea, we take the planar embedding of K_4 as an example. As shown in Figure 6.1, since the surface is orientable, we can choose a consistent orientation of the face boundaries. This partitions the arcs of K_4 into four groups $\{f_0, f_1, f_2, f_3\}$, called the *facial walks*. Meanwhile, the arcs can be partitioned into another four groups, each having the same tail. We represent these two partitions by the incidence matrices in Equation (6.1.1).

$$
M = \begin{array}{c} \\ (0,1) \\ (0,2) \\ (0,3) \\ (1,0) \\ (1,2) \\ (1,3) \\ (2,0) \\ (2,1) \\ (2,3) \\ (3,0) \\ (3,1) \\ (3,2) \end{array}
\begin{array}{cccc} f_0 & f_1 & f_2 & f_3 \\ \left(\begin{array}{cccc} 1 & 0 & 0 & 0 \\ 0 & 0 & 1 & 0 \\ 0 & 0 & 0 & 1 \\ 0 & 0 & 0 & 1 \\ 1 & 0 & 0 & 0 \\ 0 & 1 & 0 & 0 \\ 1 & 0 & 0 & 0 \\ 0 & 1 & 0 & 0 \\ 0 & 0 & 1 & 0 \\ 0 & 0 & 1 & 0 \\ 0 & 0 & 0 & 1 \\ 0 & 1 & 0 & 0 \end{array}\right) \end{array}
\qquad
N = \begin{array}{c} \\ (0,1) \\ (0,2) \\ (0,3) \\ (1,0) \\ (1,2) \\ (1,3) \\ (2,0) \\ (2,1) \\ (2,3) \\ (3,0) \\ (3,1) \\ (3,2) \end{array}
\begin{array}{cccc} 0 & 1 & 2 & 3 \\ \left(\begin{array}{cccc} 1 & 0 & 0 & 0 \\ 1 & 0 & 0 & 0 \\ 1 & 0 & 0 & 0 \\ 0 & 1 & 0 & 0 \\ 0 & 1 & 0 & 0 \\ 0 & 1 & 0 & 0 \\ 0 & 0 & 1 & 0 \\ 0 & 0 & 1 & 0 \\ 0 & 0 & 1 & 0 \\ 0 & 0 & 0 & 1 \\ 0 & 0 & 0 & 1 \\ 0 & 0 & 0 & 1 \end{array}\right) \end{array}
$$

$$(6.1.1)$$

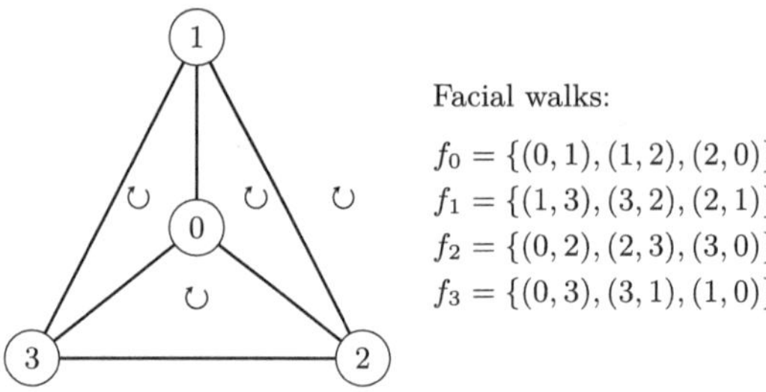

Figure 6.1   A planar embedding of K_4

If $\widehat{M}$ is the matrix obtained from M by scaling each column to a unit vector, then $\widehat{M}\,\widehat{M}^T$ is the orthogonal projection onto $\mathrm{col}(M)$, and so

$$2\widehat{M}\,\widehat{M}^T - I$$

is the reflection about $\mathrm{col}(M)$. Similarly, if $\widehat{N}$ is the normalized arc-tail incidence matrix, then

$$2\widehat{N}\,\widehat{N}^T - I$$

is the reflection about $\mathrm{col}(N)$. Now

$$U := (2\widehat{M}\,\widehat{M}^T - I)(2\widehat{N}\,\widehat{N}^T - I)$$

is a unitary matrix, which serves as the transition matrix of our discrete quantum walk.

We can easily generalize the above construction to any orientable embedding of a graph X; such a walk will be called a *vertex-face walk*. While this model has never been studied, there are search algorithms that effectively use the vertex-face walk of a toroidal embedding of $C_n \square C_n$ [56, 25, 5]. Section 6.8 will discuss this connection in more detail.

The transition matrix U of the previous example has an interesting expression: $U = \exp(tS)$, where $t \in \mathbb{R}$ and S is the skew-adjacency matrix of an oriented graph, as shown in Figure 7.1. Thus, U can be seen as the transition matrix of a continuous quantum walk, evaluated at time t. We would like to characterize vertex-face walks that are connected to continuous quantum walks in this way.

Our approach is spectral. We first list some basic properties of the transition matrix of a vertex-face walk. Then we establish a spectral correspondence between the transition matrix and the vertex-face incidence matrix. Using the incidence graph, we derive a formula for the principal logarithm of U^2. We then explore necessary conditions and sufficient conditions for the underlying digraph of this logarithm to be an oriented graph, and find interesting

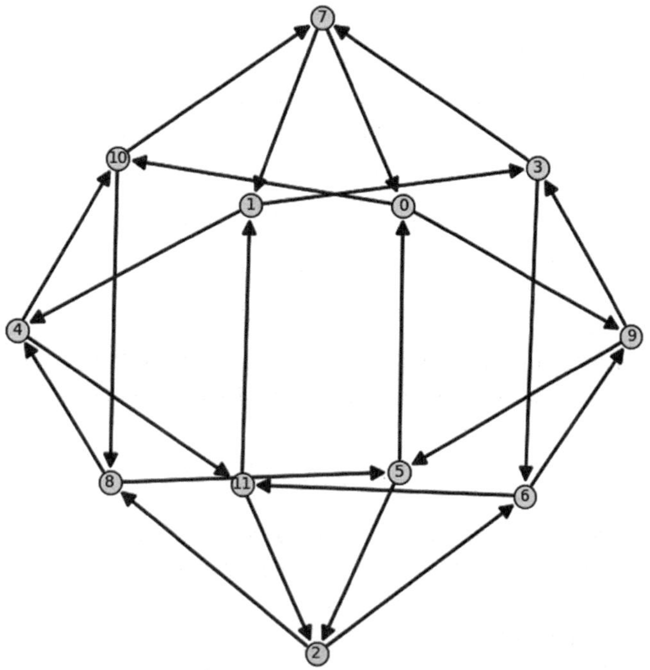

Figure 6.2 *H*-digraph of the planar embedding of K_4

connections to partial geometric designs. We also investigate properties of vertex-face walks on the covers of a graph. Finally, we note that some vertex-face walks are reluctant to leave their initial state, which is in sharp contrast to classical random walks.

6.2 Model

An embedding is *circular* if every face is bounded by a cycle. In this section, we generalize the example in Section 6.1 to a vertex-face walk on any circular orientable embedding.

Let X be a graph, and $\mathcal{M}$ an embedding of X on some orientable surface. Consider a consistent orientation of the faces, that is, for each edge e shared by two faces f and h, the direction e receives in f is opposite to the direction it receives in h. Given such an orientation, every arc belongs to exactly one face; let M be the associated arc-face incidence matrix. We also partition the arcs according to their tails, and let N be the associated arc-tail incidence matrix. Denote the normalized versions of M and N by $\widehat{M}$ and $\widehat{N}$. The unitary matrix

$$U := (2\widehat{M}\,\widehat{M}^T - I)(2\widehat{N}\,\widehat{N}^T - I)$$

is the transition matrix of a vertex-face walk for $\mathcal{M}$.

Although U depends on the consistent orientation, there are only two choices – reversing all the arcs in the facial walks of one orientation produces the other. Let R be the permutation matrix that swaps arc (u, v) with arc (v, u). If

$$(2\widehat{M}\,\widehat{M}^T - I)(2\widehat{N}\,\widehat{N}^T - I)$$

is the transition matrix relative to the "clockwise" orientation, then

$$R(2\widehat{M}\,\widehat{M}^T - I)R(2\widehat{N}\,\widehat{N}^T - I)$$

is the transition matrix relative to the "counterclockwise" orientation. In this chapter, we will not specify the orientation when proving properties of U, as our results are independent of the choice.

The following observation on duality is immediate.

6.2.1 Lemma. *If U is the vertex-face transition matrix for $\mathcal{M}$, then U^T is the vertex-face transition matrix for the dual embedding of $\mathcal{M}$.* $\qquad\square$

Define two matrices

$$P := \widehat{M}\,\widehat{M}^T, \quad Q := \widehat{N}\,\widehat{N}^T.$$

Note that P is the projection onto vectors that are constant on each facial walk, and Q is the projection onto vectors that are constant on arcs leaving each vertex. Let $\mathbf{1}$ denote the all-ones vector. The projections P and Q satisfy the following properties.

6.2.2 Lemma. *For any arc (u, v), let f_{uv} denote the facial walk using (u, v). For any two faces f and h, let $f \cap h$ denote the set of vertices used by both f and h.*

(i) *The projections P and Q are doubly stochastic, and so*

$$U\mathbf{1} = U^T\mathbf{1} = \mathbf{1}.$$

(ii) *For two arcs (a, b) and (u, v),*

$$P_{(a,b),(u,v)} = \begin{cases} \dfrac{1}{\deg(f_{uv})}, & \text{if } f_{ab} = f_{uv}. \\ 0, & \text{otherwise.} \end{cases}$$

and

$$Q_{(a,b),(u,v)} = \begin{cases} \dfrac{1}{\deg(u)}, & \text{if } a = u. \\ 0, & \text{otherwise.} \end{cases}$$

(iii) For two arcs (a, b) and (u, v),

$$(PQ)_{(a,b),(u,v)} = (QP)_{(u,v),(a,b)} = \begin{cases} \dfrac{1}{\deg(u)\deg(f_{ab})}, & \text{if } u \in f_{ab}. \\ 0, & \text{otherwise.} \end{cases}$$

(iv) For two faces f and h,

$$(\widehat{M}^T Q \widehat{M})_{f,h} = \frac{1}{\sqrt{\deg(f)\deg(h)}} \sum_{u \in f \cap h} \frac{1}{\deg(u)}.$$

For two vertices u and v,

$$(\widehat{N}^T P \widehat{N})_{u,v} = \frac{1}{\sqrt{\deg(u)\deg(v)}} \sum_{f:u,v \in f} \frac{1}{\deg(f)}.$$

Proof We prove the first parts of (iii) and (iv); the other statements follow similarly. Note that $M_{(a,b),f} \neq 0$ if and only if $f = f_{ab}$, $(M^T N)_{f,w} \neq 0$ if and only if w is contained in f, and $N^T_{w,(u,v)} \neq 0$ if and only if $w = u$. Hence $(PQ)_{(a,b),(u,v)} \neq 0$ if and only if $u \in f_{ab}$. This proves (iii). For (iv), simply note that $(M^T N N^T M)_{f,h}$ counts the vertices that appear in both faces f and h. $\qquad\square$

The above lemma allows us to write out the entries of U explicitly. Moreover, if either X or its dual graph is regular, we have a simple expression for $\mathrm{tr}(U)$.

6.2.3 Lemma. *Suppose the circular orientable embedding of X has n vertices, ℓ edges, and s faces. If either X or its dual graph is regular, then*

$$\mathrm{tr}(U) = 2\left(\frac{ns}{\ell} - (n + s - \ell)\right).$$

Proof We have

$$U = (2P - I)(2Q - I).$$

where P and Q are projections. From (iii) in Lemma 6.2.2 we see that

$$\mathrm{tr}(PQ) = \sum_{(u,v)} \frac{1}{\deg(u)} \frac{1}{\deg(f_{uv})}$$

$$= \sum_{f} \frac{1}{\deg(f)} \sum_{u \in f} \frac{1}{\deg(u)}.$$

If X is d-regular, then

$$\mathrm{tr}(PQ) = \frac{s}{d} = \frac{ns}{2\ell}.$$

Hence

$$\begin{aligned}
\operatorname{tr}(U) &= 4\operatorname{tr}(PQ) - 2\operatorname{tr}(P) - 2\operatorname{tr}(Q) - \operatorname{tr}(I) \\
&= 2\frac{ns}{\ell} - 2(\operatorname{rk}(P) + \operatorname{rk}(Q) - 2\ell) \\
&= 2\left(\frac{ns}{\ell} - (n + s - \ell)\right).
\end{aligned}$$

The case where the dual graph is regular follows from $\operatorname{tr}(U) = \operatorname{tr}(U^T)$. $\qquad\square$

A quantum walk is called *reducible* if U is a permutation similar to some block-diagonal matrix, and *irreducible* otherwise. The following result shows that for a connected graph, any vertex-face walk is irreducible, regardless of the embedding.

6.2.4 Lemma. *Let $\mathcal{M}$ be an orientable embedding of a connected graph X. Let π_1 and π_2 be the arc-face partition and the arc-tail partition of $\mathcal{M}$. Then $\pi_1 \wedge \pi_2$ is the discrete partition, and $\pi_1 \vee \pi_2$ is the trivial partition.*

Proof First of all, since every face is bounded by a cycle, no two arcs sharing the tail are contained in the same facial walk, so $\pi_1 \wedge \pi_2$ is the discrete partition. Next, since X is connected, between any two vertices v_0 and v_k there is a path, say

$$v_0, \ldots, v_k.$$

Consider the first two arcs (v_0, v_1) and (v_1, v_2). If they belong to the same facial walk, then they are in the same class of $\pi_1 \vee \pi_2$. Otherwise, there is an arc (v_1, w_1) that is in the same facial walk as (v_0, v_1). Thus, all outgoing arcs of v_1, including (v_1, v_2), are in the same class of $\pi_1 \vee \pi_2$ as (v_0, v_1). Proceeding in this fashion, we see that all arcs in the path belong to the same class of $\pi_1 \vee \pi_2$. $\qquad\square$

6.3 Spectral Decomposition

Since the transition matrix U is a product of two reflections, we can apply the results in Chapter 2 to compute its spectral decomposition. Suppose the embedding $\mathcal{M}$ has n vertices, ℓ edges, and s faces. Let g be the genus of the orientable surface. The readers may verify the following theorems on the eigenspaces of U.

6.3.1 Theorem. *The 1-eigenspace of U is*

$$(\operatorname{col}(M) \cap \operatorname{col}(N)) \oplus (\ker(M^T) \cap \ker(N^T))$$

with dimension $\ell + 2g$. Moreover, the first subspace is simply

$$\operatorname{col}(M) \cap \operatorname{col}(N) = \operatorname{span}\{\mathbf{1}\}. \qquad \square$$

To characterize the remaining eigenspaces for U, we introduce a few more incidence matrices. A vertex is incident to a face if it is incident to an edge that is contained in the face. Let B, C, and D be the vertex-edge incidence matrix, the vertex-face incidence matrix, and the face-edge incidence matrix, respectively. Since every face is bounded by a cycle, we have the following two expressions for C.

6.3.2 Lemma. *The vertex-face incidence matrix C satisfies*

$$C = \frac{1}{2}BD^T = N^T M. \qquad \square$$

We also define

$$\widehat{C} := \widehat{N}^T \widehat{M},$$

and call it the *normalized vertex-face incidence matrix*. All other eigenspaces for U are determined by $\widehat{C}$.

6.3.3 Theorem. *The (-1)-eigenspace for U is*

$$\widehat{M}\ker(\widehat{C}) \oplus \widehat{N}^T \ker(\widehat{C}^T)$$

with dimension

$$n + s - 2\operatorname{rk}(C). \qquad \square$$

By interlacing, the eigenvalues of $\widehat{C}\,\widehat{C}^T$ lie in $[0, 1]$. The following theorem shows how the eigenspaces for $\widehat{C}\widehat{C}^T$ with eigenvalues in $(0, 1)$ give rise to eigenspaces for U with non-real eigenvalues.

6.3.4 Theorem. *The multiplicities of the non-real eigenvalues of U sum to $2\operatorname{rk}(C) - 2$. Let $\mu \in (0, 1)$ be an eigenvalue of $\widehat{C}\,\widehat{C}^T$. Choose θ with*

$$\cos(\theta) = 2\mu - 1.$$

The map

$$y \mapsto (\cos(\theta) + 1)\widehat{N}y - (e^{i\theta} + 1)\widehat{M}\,\widehat{C}^T y$$

is an isomorphism from the μ-eigenspace of $\widehat{C}\,\widehat{C}^T$ to the $e^{i\theta}$-eigenspace of U, and the map

$$y \mapsto (\cos(\theta) + 1)\widehat{N}y - (e^{-i\theta} + 1)\widehat{M}\,\widehat{C}^T y$$

is an isomorphism from the μ-eigenspace of $\widehat{C}\,\widehat{C}^T$ to the $e^{-i\theta}$-eigenspace of U. $\qquad\square$

After normalization, we obtain an explicit formula for the eigenprojection of each non-real eigenvalue of U.

6.3.5 Corollary. *Let $\mu \in (0,1)$ be an eigenvalue of $\widehat{C}\,\widehat{C}^T$. Choose θ such that $\cos(\theta) = 2\mu - 1$. Let E_μ be the orthogonal projection onto the μ-eigenspace of $\widehat{C}\,\widehat{C}^T$. Set*

$$W := \widehat{N}E_\mu\widehat{N}^T.$$

Then the $e^{i\theta}$-eigenprojection of U is

$$\frac{1}{\sin^2(\theta)}\left((\cos(\theta)+1)W - (e^{i\theta}+1)PW - (e^{-i\theta}+1)WP + 2PWP\right),$$

and the $e^{-i\theta}$-eigenprojection of U is

$$\frac{1}{\sin^2(\theta)}\left((\cos(\theta)+1)W - (e^{-i\theta}+1)PW - (e^{i\theta}+1)WP + 2PWP\right). \qquad\square$$

6.4 Hamiltonian

A *Hamiltonian* of a unitary matrix V is a Hermitian matrix H such that $V = \exp(iH)$. Given the spectral decomposition

$$V = \sum_r \alpha_r F_r,$$

we may take

$$H = -i\sum_r \log(\alpha_r)F_r,$$

for some value of $\log(\alpha_r)$. If, in addition, for each r the angle satisfies

$$-\pi < -i\log(\alpha_r) \le \pi,$$

then H is called the *principal Hamiltonian of V*. In this sense, every unitary matrix can be viewed as the transition matrix of a continuous quantum walk on the underlying digraph of its principal Hamiltonian.

We study the principal Hamiltonian of U^2, where U is the transition matrix of a vertex-face walk. The spectral machinery we developed in the previous section reveals a close connection between H and the bipartite vertex-face incidence graph. For simplicity, we will focus on circular orientable embeddings where both X and the dual graph are regular; if each vertex has d neighbours and each face uses k vertices, we say the embedding *has type* (k, d).

6.4.1 Theorem. *Let $\mathcal{M}$ be a circular orientable embedding of type (k, d). Let M and N be the arc-face incidence matrix and the arc-tail incidence matrix, respectively. Let U be the transition matrix of a vertex-face walk for $\mathcal{M}$. Let A be the adjacency matrix of the vertex-face incidence graph, with spectral decomposition*

$$A = \sum_\lambda \lambda G_\lambda.$$

Then each eigenvalue λ of A gives rise to some eigenvalue $e^{\pm i\theta}$ of U in the following way:

$$\frac{|\lambda|}{\sqrt{dk}} = \cos\left(\frac{\theta}{2}\right).$$

Moreover, $U^2 = \exp(iH)$, where

$$H = 4 \begin{pmatrix} N & iM \end{pmatrix} \left(\sum_{\lambda \notin \{0, \pm\sqrt{dk}\}} \frac{\lambda \arccos(|\lambda|/\sqrt{dk})}{|\lambda|\sqrt{dk - \lambda^2}} G_r \right) \begin{pmatrix} N^T \\ -iM^T \end{pmatrix}.$$

Proof As before, let C be the vertex-face incidence matrix. Then

$$A = \begin{pmatrix} 0 & C \\ C^T & 0 \end{pmatrix}.$$

Note that the eigenvalues of A are symmetric about zero and bounded in absolute value by $\sqrt{dk}$. By Theorems 6.3.1, 6.3.3, and 6.3.4, each eigenvalue λ of A determines the real part of some eigenvalue $e^{\pm i\theta}$ of U:

$$\cos(\theta) = \frac{2\lambda^2}{dk} - 1,$$

that is,

$$\frac{|\lambda|}{\sqrt{dk}} = \cos\left(\frac{\theta}{2}\right).$$

Moreover, for $\lambda \neq 0$, the λ-eigenprojection for A is

$$G_\lambda = \frac{1}{2} \begin{pmatrix} E_{\lambda^2} & \frac{1}{\lambda} E_{\lambda^2} C \\ \frac{1}{\lambda} C^T E_{\lambda^2} & \frac{1}{\lambda^2} C^T E_{\lambda^2} C \end{pmatrix},$$

where E_{λ^2} is the λ^2-eigenprojection for CC^T. Since U has real entries, its spectrum is closed under complex conjugation. Hence by Corollary 6.3.5,

$$H = -4i \sum_{0<\lambda<\sqrt{dk}} \frac{\arccos(\lambda/\sqrt{dk})}{\lambda\sqrt{dk-\lambda^2}}(NE_{\lambda^2}CM^T - MC^T E_{\lambda^2}N^T).$$

Combining this with the fact that

$$G_\lambda - G_{-\lambda} = \begin{pmatrix} 0 & \frac{1}{\lambda}E_{\lambda^2}C \\ \frac{1}{\lambda}C^T E_{\lambda^2} & 0 \end{pmatrix}$$

yields the formula for H. $\qquad\square$

The above result implies that H is the block sum of

$$\begin{pmatrix} N & 0 \\ 0 & iM \end{pmatrix} \phi(A) \begin{pmatrix} N^T & 0 \\ 0 & -iM^T \end{pmatrix}$$

for some odd polynomial ϕ.

6.5 H-Digraph

Given an embedding $\mathcal{M}$ and the vertex-face transition matrix U, we will refer to the underlying digraph of the principal Hamiltonian H of U^2 as the H-*digraph* of $\mathcal{M}$. By Theorem 6.4.1, iH is skew-symmetric, so the H-digraph is a weighted oriented graph.

An embedding $\mathcal{M}$ is *orientably regular* if its orientation-preserving automorphism group acts regularly on the arcs. Using the decomposition

$$C = \frac{1}{2}BD^T = N^T M,$$

where B, C, D, M, and N are the vertex-edge, vertex-face, face-edge, arc-face, and arc-tail incidence matrices, we obtain the following.

6.5.1 Theorem. *Let $\mathcal{M}$ be a circular orientable embedding of type (k, d). If $\mathcal{M}$ is orientably regular, then the vertex-face incidence graph is edge-transitive, and the H-digraph is vertex-transitive.* $\qquad\square$

In general, we should expect the H-digraph to be dense with many different weights; however, there are cases where it is sparse and unweighted (up to scaling). An example is given in Figure 7.1. In this section, we study circular orientable embeddings of type (k, d) whose H-digraphs are oriented graphs.

We first give a necessary condition on the eigenvalues of the vertex-face incidence graph. This is a direct consequence of a result by the first author on real state transfer [29], which we summarize below.

6.5.2 Theorem. *Let H be a Hermitian matrix with algebraic entries. Suppose for some real number t the entries of $\exp(itH)$ are algebraic. Then the ratio of any two non-zero eigenvalues of H are rational. Moreover, if iH has integer entries, then there is a square-free integer Δ such that all eigenvalues of H are in $\mathbb{Z}[\sqrt{\Delta}]$.* $\square$

The ratio condition in Theorem 6.5.2 is particularly useful in characterizing state transfer in continuous quantum walks on graphs (see, for example, [27]) and oriented graphs ([29, 51]). Here, we present its application to discrete quantum walks.

6.5.3 Theorem. *Let U be a vertex-face transition matrix for a circular orientable embedding of type (k, d). Let A be the adjacency matrix of the vertex-face incidence graph. Suppose $U^2 = \exp(tS)$ for some real number t and skew-adjacency matrix S. Then the following hold.*

(i) *There is a square-free integer Δ such that all eigenvalues of S are in $\mathbb{Z}[\sqrt{-\Delta}]$.*

(ii) *If λ_r and λ_s are two eigenvalues of A that are distinct from $\{0, \pm\sqrt{dk}\}$, then*

$$\frac{\arccos(|\lambda_r|/\sqrt{dk})}{\arccos(|\lambda_s|/\sqrt{dk})} \in \mathbb{Q}.$$

Proof (i) follows from Theorem 6.4.1 since U has rational entries and S has integer entries. For (ii), recall that λ_r and λ_s determine non-real eigenvalues $e^{\pm i\theta_r}$ and $e^{\pm i\theta_s}$ of U by

$$\frac{|\lambda_r|}{\sqrt{dk}} = \cos\left(\frac{\theta_r}{2}\right), \qquad \frac{|\lambda_s|}{\sqrt{dk}} = \cos\left(\frac{\theta_s}{2}\right).$$

If $0 < \theta_r, \theta_s < \pi$, then θ_r/θ_s equals the ratio of two eigenvalues of S, which must be rational. $\square$

The above condition is satisfied when U^2 has exactly three eigenvalues. In fact, the H digraph of such U^2 is guaranteed to be an oriented graph.

6.5.4 Theorem. *Let $\mathcal{M}$ be a circular orientable embedding of type (k, d) of a graph X. Let U be a vertex-face transition matrix for $\mathcal{M}$. Then*

$$U^2 = \exp(\gamma(U - U^T))$$

for some real number γ if and only if the vertex-face incidence graph has four or five distinct eigenvalues. Moreover,

$$\frac{dk}{4}(U^T - U)$$

is the skew-adjacency matrix of some oriented graph on the arcs of X, and the degree of (a, b) is

$$dk - \sum_{u \in f_{ab}} \beta(a, u),$$

where f_{ab} denotes the unique face using the arc (a, b), and $\beta(a, u)$ denotes the number of faces containing both a and u.

Proof Let

$$U = \sum_r \alpha_r F_r$$

be the spectral decomposition of U. Then

$$U^2 = \exp(\gamma(U - U^T))$$

holds if and only if

$$\sum_r \alpha_r^2 F_r = \sum_r e^{\gamma(\alpha_r - \alpha_r^{-1})} F_r,$$

that is, for each non-real eigenvalue $\alpha_r = e^{i\theta_r}$ of U,

$$e^{2i\theta} = e^{2\gamma \sin(\theta)}.$$

Since $\sin(x)/x$ is monotone when $0 < x < \pi$, the above holds for some γ if and only if U^2 has exactly three eigenvalues, or equivalently, the vertex-face incidence graph has exactly four or five eigenvalues.

    Let M and N be the arc-face and arc-tail incidence matrices, respectively. Recall that

$$U = \left(\frac{2}{k}MM^T - I\right)\left(\frac{2}{d}NN^T - I\right).$$

Thus,

$$U - U^T = \frac{4}{dk}(MM^T NN^T - NN^T MM^T).$$

Let $S = MM^T NN^T - NN^T MM^T$. By Lemma 6.2.2 (iii),

$$S_{(a,b),(u,v)} = \begin{cases} 1, & \text{if } u \in f_{ab} \text{ and } a \notin f_{uv}, \\ -1, & \text{if } a \in f_{uv} \text{ and } u \notin f_{ab}, \\ 0, & \text{otherwise.} \end{cases}$$

Therefore S is the skew-adjacency matrix of some oriented graph. Moreover, for each $u \in f_{ab}$, there is a bijection between the neighbours of v such that $a \in f_{uv}$ and the faces containing both a and u. Hence the degree of the oriented graph is

$$\sum_{u \in f_{ab}} (d - \beta(a,u)) = dk - \sum_{u \in f_{ab}} \beta(a,u). \qquad \square$$

Let C be the vertex-face incidence matrix of the embedding in Theorem 6.5.4. We see that $U^2 = \exp(\gamma(U - U^T))$ if and only if C has exactly two non-zero singular values. Combinatorial designs with two non-zero singular eigenvalues were studied by van Dam and Spence [68, 69]. In particular, they showed that a point-d-regular and block-k-regular design with two non-zero singular values is a *partial geometric design* with parameters (d, k, t, c), originally introduced by Bose and colleagues [13], where for each point-block pair (p, B), the number of incident point-block pairs

$$|\{(p', B') : p' \neq p, B' \neq B, p' \in B, p \in B'\}|$$

equals c or t, depending on whether p is in B or not. Below we include a proof.

6.5.5 Theorem. *Let C be an incidence matrix with $C\mathbf{1} = d\mathbf{1}$ and $C^T\mathbf{1} = k\mathbf{1}$. Suppose C has exactly two non-zero singular values. Then C is the incidence matrix of a partial geometric design.*

Proof Clearly, dk is an eigenvalue of CC^T with eigenprojection J/n. Let μ be the other non-zero eigenvalue of CC^T. Let E_0 be the projection onto the kernel of CC^T. We have

$$CC^T = dk \left(\frac{1}{n}J\right) + \mu \left(I - \frac{1}{n}J - E_0\right) = \mu I + \frac{dk - \mu}{n}J - \mu E_0.$$

Thus,

$$CC^TC = \mu C + \frac{k(dk - \mu)}{n}J.$$

Note that $(CC^TC)_{p,B}$ counts all pairs (p', B') with $p' \in B$ and $p \in B'$. Hence the incidence structure is a

$$\left(d, k, \frac{k(dk - \mu)}{n}, \frac{k(dk - \mu)}{n} + \mu + 1 - d - k\right)$$

partial geometric design. $\qquad \square$

We briefly discuss embeddings that realize partial geometric designs.

If CC^T is invertible, then a trace argument shows that the vertex-face incidence structure is a *balanced incomplete block design (BIBD)*, or a 2-*design*, with parameters $(n, k, d(k-1)/(n-1))$, that is, a point-regular and block-regular design where every two distinct points lie in $d(k-1)/(n-1)$ blocks.

The study of connections between 2-designs and graph embeddings dates back to 1897 [41], when Heffter constructed 2-$(n, 3, 2)$ designs using certain triangular embeddings of K_n. However, building these triangular embeddings themselves remained a challenging task, until Ringel [61], Gustin [39], and Terry and colleagues [66] provided solutions to all admissible n, that is, $n \equiv 0, 3, 4, 7 \pmod{12}$.

The self-dual circular embeddings of K_n, on the other hand, yield a family of 2-$(n, n-1, n-2)$ designs. While self-dual embeddings of K_n exist if and only if $n \equiv 0, 1 \pmod{4}$ [72], the circular ones are only known when n is a prime power. For the constructions, see Biggs [12]; we remark that these are all orientably regular embeddings.

Embeddings with singular CC^T and $C^T C$ may be related to other designs. A *two-class partially balanced incomplete block design (PBIBD)* with parameters $(n, k; \lambda_1, \lambda_2)$ is a point-regular, block-regular design whose incidence matrix C satisfies

$$CC^T = dI + \lambda_1 A + \lambda_2 (J - I - A),$$

where A is the adjacency matrix of a strongly regular graph. A PBIBD has at most three non-zero singular values; those with two non-zero singular values are precisely partial geometric PBIBDs, and they are usually referred to as *special PBIBDs* [14].

Every triangular embedding of a strongly regular graph on n vertices determines a $(n, 3; 2, 0)$-PBIBD. In [57], Petroelje gave a construction for orientable triangular embeddings of $K_{n,n,n}$, which yield $(3n, 3; 2, 0)$-special PBIBDs.

6.6 Covers

In this section, we consider a covering construction that preserves nice properties of vertex-face walks. For a detailed discussion on covers, see Chapter 5.

Given an orientable embedding $\mathcal{M}_X$ of X, and a covering map ψ from a connected graph Y to X, we define an orientable embedding $\mathcal{M}_Y$ of Y by specifying its facial walks. Let W be a facial walk of $\mathcal{M}_X$ starting at vertex u. Clearly, the preimage $\psi^{-1}(W)$ consists of walks starting and ending in the fiber $\psi^{-1}(u)$, and each arc of Y appears in at most one of these walks. Then, the facial walks of $\mathcal{M}_Y$ are exactly the closed walks in the preimages of the facial walks of $\mathcal{M}_X$. In the previous example, the planar embedding of K_4 gives rise to an embedding of the cube on the torus, with four faces each of length 6.

We will focus on a special type of cover, known as the voltage graphs. We call $\mathcal{M}_Y$ a *voltage embedding* if Y is a voltage graph of X, and $\mathcal{M}_Y$ is obtained from $\mathcal{M}_X$ by the above lifting method.

The next result shows that the transition matrix of $\mathcal{M}_X$ is a block sum of the transition matrix of $\mathcal{M}_Y$, and consequently, the H-digraph of $\mathcal{M}_X$ is a quotient digraph of $\mathcal{M}_Y$. To prove it, we need the concept of row and column equitable partitions, which were introduced by the first author [26, chapter 12]. Let A be a matrix over $\mathbb{C}$. Let σ and ρ be the partition of the columns and rows of A, and let K and L be their respective characteristic matrices. The pair (ρ, σ) is *column equitable* if $\mathrm{col}(AK) \subseteq \mathrm{col}(L)$, *row equitable* if $\mathrm{col}(A^*L) \subseteq \mathrm{col}(K)$, and *equitable* if it is both column and row equitable.

6.6.1 Theorem. *Let $\mathcal{M}_X$ be a circular orientable embedding of X. Let Y be a voltage graph of X, and $\mathcal{M}_Y$ the associated voltage embedding. Let ρ be the partition of the arcs of Y, where each class is the preimage of some arc of X. Let $\widehat{L}$ be its normalized incidence matrix of ρ. If U_X and U_Y are the vertex-face transition matrices for $\mathcal{M}_X$ and $\mathcal{M}_Y$, then*

$$U_X = \widehat{L}^T U_Y \widehat{L}.$$

Consequently, the H-digraph of $\mathcal{M}_X$ is a quotient digraph of the H-digraph of $\mathcal{M}_Y$.

Proof Let $\widehat{M}_X, \widehat{M}_Y, \widehat{N}_X, \widehat{N}_Y$ be the arc-face incidence matrices and arc-tail incidence matrices of the embeddings of X and Y, respectively. We have

$$U_Y = (2\widehat{M}_Y \widehat{M}_Y^T - I)(2\widehat{N}_Y \widehat{N}_Y^T - I).$$

Let σ be the partition of the vertices of Y into fibers, with normalized incidence matrix $\widehat{K}$. It is not hard to verify that

$$\widehat{N}_Y \widehat{K} = \widehat{L} \widehat{N}_X$$

and

$$\widehat{N}_Y^T \widehat{L} = \widehat{K} \widehat{N}_X^T.$$

Thus, (ρ, σ) is an equitable partition of $\widehat{N}_Y$. It follows that

$$\widehat{N}_Y \widehat{K} \widehat{K}^T = \widehat{L} \widehat{L}^T \widehat{N}_Y. \tag{6.6.1}$$

Since

$$\widehat{N}_X = \widehat{L}^T \widehat{N}_Y \widehat{K},$$

the projection onto $\mathrm{col}(\widehat{N}_X)$ can be written as

$$\begin{aligned}
\widehat{N}_X \widehat{N}_X^T &= \widehat{L}^T (\widehat{N}_Y \widehat{K} \widehat{K}^T) \widehat{N}_Y^T \widehat{L} \\
&= \widehat{L}^T (\widehat{L} \widehat{L}^T \widehat{N}_Y) \widehat{N}_Y^T \widehat{L} \\
&= \widehat{L}^T \widehat{N}_Y \widehat{N}_Y^T \widehat{L}.
\end{aligned}$$

Applying a similar argument to the preimages of facial walks, we can show that

$$\widehat{M}_X \widehat{M}_X^T = \widehat{L}^T \widehat{M}_Y \widehat{M}_Y^T \widehat{L}.$$

Thus,

$$U_X = \widehat{L}^T (2\widehat{M}_Y \widehat{M}_Y^T - I) \widehat{L} \widehat{L}^T (2\widehat{N}_Y \widehat{N}_Y^T - I) \widehat{L}. \tag{6.6.2}$$

Finally, from Equation (6.6.1) we see that

$$\widehat{L} \widehat{L}^T \widehat{N}_Y \widehat{N}_Y^T = \widehat{N}_Y \widehat{K} \widehat{K}^T \widehat{N}_Y^T,$$

which is a symmetric matrix, so $\widehat{L} \widehat{L}^T$ commutes with $\widehat{N}_Y \widehat{N}_Y^T$. Therefore, Equation (6.6.2) reduces to

$$U_X = \widehat{L}^T U_Y \widehat{L}. \qquad \square$$

Conversely, we may "lift" nice properties of $\mathcal{M}_X$ when taking a voltage embedding, using the following simple technique. Let C be the incidence matrix of a design. Construct a new design with incidence matrix $C \otimes \mathbf{1}$ by duplicating the points and preserving the incidence relation. If C is a partial geometric design, then so is $C \otimes \mathbf{1}$.

6.6.2 Theorem. *Let $\mathcal{M}_X$ be a circular orientable embedding of X. Suppose its vertex-face incidence structure is a partial geometric design. Let $Y = X^\phi$ be a voltage graph of order r, and $\mathcal{M}_Y$ the associated voltage embedding. Suppose for each facial cycle C of $\mathcal{M}_X$, the order of $\phi(C)$ is r. Then the vertex-face incidence structure of $\mathcal{M}_Y$ is also a partial geometric design.*

Proof Let ψ be the covering map. By Theorem 6.6.1, each facial cycle f_{ab} of $\mathcal{M}_X$ lifts to a unique facial cycle $\psi^{-1}(f_{ab})$ of $\mathcal{M}_Y$. Moreover, all arcs in $\psi^{-1}((a,b))$ are contained in $\psi^{-1}(f_{ab})$. $\qquad \square$

This construction yields many new embeddings whose H-digraphs are oriented graphs. For example, we have the following family based on the circular self-dual embeddings of K_n.

6.6.3 Corollary. *Let n be a power of 2. Let $\mathcal{M}_{K_n}$ be a circular self-dual embedding of K_n. Let $Y = K_n^{\phi}$ be the double cover of K_n with ϕ sending every arc to the involution. Let $\mathcal{M}_Y$ be the voltage embedding. Then the H-digraph of $\mathcal{M}_Y$ is an oriented graph.* $\qquad\square$

6.7 Sedentary Walks

One counterintuitive phenomenon in quantum walks is that the walker may be reluctant to leave its initial state. This was first observed in continuous quantum walks on K_n: for any time t, the mixing matrix

$$U_{K_n}(t) \circ \overline{U_{K_n}(t)}$$

converges to I as n goes to infinity. In [30], the first author investigated quantum walks on complete graphs, some cones and some strongly regular graphs that enjoy the same property. Following his paper, we say a sequence of discrete quantum walks, determined by transition matrices $\{U_1, U_2, \cdots\}$, is *sedentary* if for any step t, the mixing matrices $U_n^t \circ \overline{U_n^t}$ converge to I as n goes to infinity.

6.7.1 Lemma. *Let $\mathcal{M}$ be a circular orientable embedding of type (k, d). Suppose the vertex-face incidence structure is a 2-design. Then*

$$\mathrm{tr}(U^t) = nd - 2(1 - \cos(t\theta))(n - 1),$$

where

$$\cos(\theta) = \frac{2(n - k)}{k(n - 1)} - 1.$$

Proof Let C be the vertex-face incidence matrix. We have

$$CC^T = \frac{d(n - k)}{n - 1}I + \frac{d(k - 1)}{n - 1}J.$$

The eigenvalues of CC^T are dk with multiplicity 1, and $d(n - k)/(n - 1)$ with multiplicity $n - 1$. By Theorem 6.3.4, the non-real eigenvalues of U are $e^{\pm i\theta}$, each with multiplicity $n - 1$, where

$$\cos(\theta) = \frac{2(n - k)}{k(n - 1)} - 1.$$

Hence 1 is an eigenvalue of U with multiplicity $nd - 2(n - 1)$. Therefore,

$$\mathrm{tr}(U^t) = (e^{it\theta} + e^{-it\theta})(n - 1) + nd - 2(n - 1),$$

from which the result follows. $\qquad\square$

We found one family of sedentary walks from embeddings we visited before.

6.7.2 Corollary. *For each prime power n, let U_n be the vertex-face transition matrix for a self-dual orientably regular embedding of K_n. The quantum walks determined by*

$$\{U_n : n \text{ is a prime power}\}$$

form a sedentary family.

Proof Since the embedding is orientably regular, U^t has constant diagonal. By Theorem 6.7.1, each diagonal entry of $U^t \circ U^t$ is

$$\left(1 - \frac{2(1 - \cos(t\theta))}{n}\right)^2,$$

which converges to 1 as n goes to infinity. $\qquad\square$

6.8 Search

We mention a potential application of the vertex-face walks. First, let us revisit a quantum walk-based algorithm due to Patel and colleagues [56] and Falk [25]; its performance was proved to match the best-known quantum algorithms for searching a marked item on a 2-dimensional grid [5] .

We will view the 2-dimensional grid as a Cartesian square of a cycle:

$$X := C_n \square C_n.$$

Consider two partitions of $V(X)$, illustrated by the (thick) solid squares and the dashed squares in Figure 6.3. Let U_o and U_e be the reflections about the column spaces of the characteristic matrices of these two partitions, respectively. If we remove the oracle from the search algorithm proposed by [56, 25], then it is equivalent to a quantum walk with transition matrix

$$U = U_o U_e.$$

Notice that Figure 6.3 represents a self-dual embedding of $C_n \square C_n$ on the torus. In fact, it gives rise to a toroidal embedding of another graph Y, obtained by truncating the edges of X and joining the new vertices by solid and dashed edges, as shown in Figure 6.4.

Clearly, Y is isomorphic to $C_{2n} \square C_{2n}$, and the solid and dashed squares partition $V(Y)$ the same way they do in Figure 6.3. Thus, based on [5], we can construct a transition matrix U from these two partitions. On the other hand, we may think of the vertices of Y as arcs of X – the one closer to u on edge $\{u, v\}$ is the arc (u, v), with tail u. Thus, the solid squares partition the arcs based on their tails, while the dashed squares partition the arcs based on the

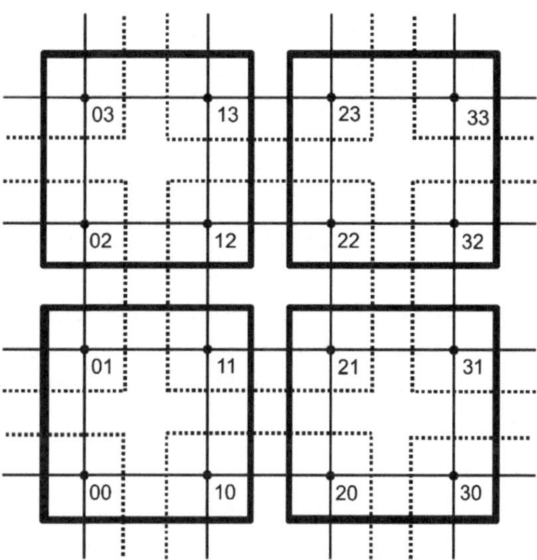

Figure 6.3 Two partitions of the vertices of $C_n \square C_n$

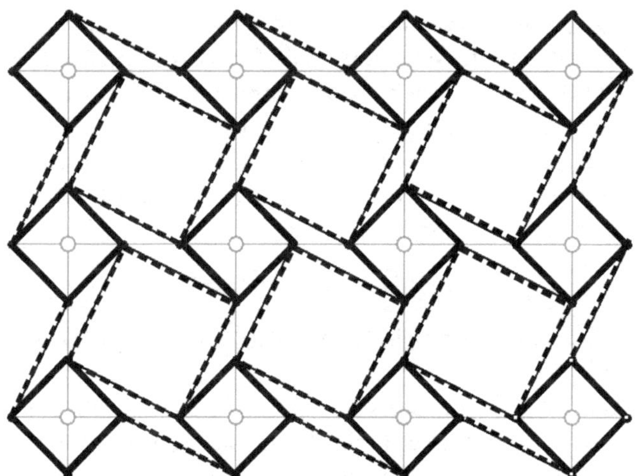

Figure 6.4 Two partitions of the arcs of $C_n \square C_n$

faces they lie in. Therefore, U is a vertex-face transition matrix for the toroidal embedding of X.

In general, given a quantum walk with transition matrix U on the arcs of a graph, we may search for a marked vertex u in the following way. Let O be the matrix that maps e_a to $-e_a$ if a is an outgoing arc of u, and fixes e_a otherwise;

this is called the *oracle*. Initialize the system to $1/\sqrt{m}$, where m is the number of arcs. Apply OU to the initial state t times. The probability of finding u after t steps is given by

$$\sum_{a \text{ has tail } u} |\langle (OU)^t \frac{1}{\sqrt{m}} \mathbf{1}, e_a \rangle|^2 .$$

We remark that each step of a vertex-face walk is equivalent to two steps of the arc-reversal walk, one on the original graph and one on the dual graph. Our computation shows that search using a vertex-face walk has a higher success probability than search using the arc-reversal walk. Of course, one reason is that the walker may move to non-adjacent arcs during each iteration of the vertex-face walk.

6.9 Notes

The definition of vertex-face walks can be extended to non-orientable embeddings through graph-encoded maps. Note that an embedding $\mathcal{M}$ with ℓ edges has 4ℓ flags. Thus, if $\mathcal{M}$ is orientable, then there are two components in the distance-2 graph of the gem, each with 2ℓ vertices. Let Y be one such component. We claim that the vertex-face walk for $\mathcal{M}$ is equivalent to a quantum walk on the vertices of Y. Let π_1 be the partition of the vertices (u, e, f) of Y based on their third coordinates f. It is not hard to see that the size of each cell in π_1 is the degree of some face. Similarly, let π_2 be the partition of $V(Y)$ based on their first coordinates u. Let $\widehat{M}$ and $\widehat{N}$ be the normalized characteristic matrices for π_1 and π_2, respectively. Then

$$(2\widehat{M}\,\widehat{M}^T - I)(2\widehat{N}\,\widehat{N}^T - I)$$

is precisely the vertex-face walk for $\mathcal{M}$ relative to one consistent orientation of the faces.

In general, let π_1 be the coarsest partition of the flags for some circular embedding $\mathcal{M}$, such that in each cell, all flags share a face, while no two flags share an edge. Similarly, let π_2 be the coarsest partition of the flags, such that in each cell, all flags share a vertex, while no two flags share an edge. Let $\widehat{M}$ and $\widehat{N}$ be the normalized characteristic matrices for π_1 and π_2, respectively. Then

$$U = (2\widehat{M}\,\widehat{M}^T - I)(2\widehat{N}\,\widehat{N}^T - I)$$

defines a quantum walk, which is reducible if and only if $\mathcal{M}$ is orientable. Now, each arc (u, v) in the underlying graph X is paired with two flags (u, e, f) and (u, e, f'), and the probability that the walker is on the arc (u, v) can be computed by summing the probabilities of her being on (u, e, f) and (u, e, f'). There are many questions we may ask about this new definition of vertex-face walks; for example, one may study the relation between the original graph X and the H-digraph, or compare the dynamics of vertex-face walks between orientable and non-orientable embeddings.

7

Shunts

In Section 1.7, we showed that Grover's walk could be expressed using a unitary matrix $U = R(I \otimes G)$. Here R is the permutation matrix that swaps each arc with its reverse, while $I \otimes G$ is a direct sum of copies of a (unitary) operation that maps the complex span of the arc starting at a vertex to itself. (The diagonal blocks are the *coins*, in this case the Grover coin.) In this chapter we study a second class of walks on the arcs of a d-regular directed graph (which means that each vertex has exactly d out-neighbours and d in-neighbours). The idea is to allow a range of choices of permutations matrices to replace R.

A *shunt* on a directed graph X is a permutation of its vertices that maps each vertex of the graph to one of its out-neighbours. A *shunt decomposition* is a collection of shunts that partition the arcs of X. Since X is d-regular, a shunt decomposition consists of exactly d shunts. A shunt decomposition of a d-regular graph can be specified by a set of permutation matrices $P_1, \ldots, P_d$ such that

$$A = P_1 + \cdots + P_d.$$

Given such a shunt decomposition, we define a permutation matrix S by

$$S = \begin{pmatrix} P_1^{-1} & & & \\ & P_2^{-1} & & \\ & & \ddots & \\ & & & P_d^{-1} \end{pmatrix}.$$

If $n = |V(X)|$, each permutation matrix P_i is $n \times n$ and thus S is $nd \times nd$. Next we choose a unitary coin C of order $d \times d$, and the transition matrix of our *shunt decomposition walk* is

$$U = S(C \otimes I_n).$$

Such walks were first defined by Aharonov et al. [1].

Our dihedral machinery from Chapter 2 can only be applied if $S^2 = I$, which generally does not hold. So in this chapter we focus on the case where the permutation matrices in the shunt decomposition commute; equivalently, when X is a Cayley digraph over an abelian group Γ. Given that every vertex receives the same coin, the spectral decomposition of U is now determined by the coin and the characters of Γ, as was originally observed by Aharonov and colleagues [1]. We apply their results to shunt-decomposition Grover walks and obtain explicit formulas for the eigenvalues and eigenvectors.

As pointed out in [1], a shunt-decomposition walk admits uniform average vertex mixing if U has distinct eigenvalues. With Grover coins, however, U will always have -1 as a non-simple eigenvalue unless $d = 2$. Therefore, previous studies on uniform average vertex mixing concentrated on cycles with more complicated coins. We show that for a shunt-decomposition Grover walk, the simple-eigenvalue condition is unnecessary, thus opening up possibilities for more examples with higher degrees. Using tools from algebraic number theory, we prove that for any prime p, a 3-regular circulant digraph over $\mathbb{Z}_p$ admits uniform average vertex mixing if and only if its connection set has trivial stabilizer in $\mathrm{Aut}(\mathbb{Z}_p)$. This provides the first infinite family of digraphs, other than cycles, that admit uniform average mixing.

Finally, we give an overview of a different approach to shunt-decomposition walks on infinite graphs, due to Ambainis and colleagues [3]. This was the first paper on shunt-decomposition models where exact analysis was carried out.

7.1 Shunt-Decomposition Walks

If a digraph admits a shunt-decomposition, then it must be regular. Using standard results in graph theory, we show that the converse is also true.

7.1.1 Lemma. *Let X be a d-regular digraph. Then X admits a shunt-decomposition.*

Proof Let A be the adjacency matrix of X. Define

$$B := \begin{pmatrix} 0 & A \\ A^T & 0 \end{pmatrix}.$$

Then B is the adjacency matrix of a d-regular bipartite graph. It is a well-known fact that every regular bipartite graph has a 1-factorization, whence

$$B = \begin{pmatrix} 0 & P_1 \\ P_1^T & 0 \end{pmatrix} + \cdots + \begin{pmatrix} 0 & P_d \\ P_d^T & 0 \end{pmatrix},$$

where $P_1, \ldots, P_d$ are permutation matrices. Therefore,

$$A(X) = P_1 + \cdots + P_d. \qquad \square$$

In the rest of this chapter, assume X is a d-regular digraph. Let C be a $d \times d$ unitary coin. As with the arc-reversal C-walk (see Section 1.7), for each vertex u, we need to specify a linear order on the neighbours of u:

$$f_u : \{1, 2, \ldots, \deg(u)\} \to \{v : u \sim v\},$$

in order to construct the coin matrix. One way to do this is to choose the linear order f_u according to a shunt-decomposition of X:

$$A(X) = P_1 + \cdots + P_d;$$

that is, for $j = 1, 2, \ldots, d$, set $f_u(j) = e_v$ where v is the unique vertex such that

$$P_j^{-1} e_u = e_v.$$

Given linear orders

$$\{f_u : u \in V(X)\},$$

the coin C sends $(u, f_u(j))$ to a superposition of all outgoing arcs of u, in which the amplitudes come from the jth column of C:

$$Ce_j = \sum_{k=1}^{d} (e_k^T Ce_j) e_k.$$

Now let

$$A(X) = P_1 + \cdots + P_d$$

be a shunt-decomposition of X, and let

$$S = \begin{pmatrix} P_1^{-1} & & & \\ & P_2^{-1} & & \\ & & \ddots & \\ & & & P_d^{-1} \end{pmatrix}.$$

The ordering of the rows and columns of S defines a set of linear orders $\{f_u : u \in V(X)\}$. Choose a $d \times d$ unitary coin C, and assign it to each vertex u according to f_u. Then the coin matrix can be written as $C \otimes I$, and the transition matrix as

$$U = S(C \otimes I).$$

Note that $S^n = I$. In general, S and $C \otimes I$ do not commute, and the spectral decomposition of U could be very hard to derive. However, if $P_1, \ldots, P_d$ have a common eigenvector χ, then we can use χ to construct an eigenvector for U. This observation is due to Aharonov et al. [1].

7.1.2 Lemma. *Let X be a digraph with shunt-decomposition*

$$A(X) = P_1 + \cdots + P_d.$$

Let U be the transition matrix of a shunt-decomposition C-walk on X. Let χ be a common eigenvector for the shunts $P_1, \ldots, P_d$ with eigenvalues $\lambda_1, \ldots, \lambda_d$, respectively. Then $y \otimes \chi$ is an eigenvector for U with eigenvalue α if and only if y is an eigenvector for

$$\begin{pmatrix} \lambda_1^{-1} & & & \\ & \lambda_2^{-1} & & \\ & & \ddots & \\ & & & \lambda_d^{-1} \end{pmatrix} C$$

with eigenvalue α.

Proof We have

$$U(y \otimes \chi) = \begin{pmatrix} P_1^{-1} & & & \\ & P_2^{-1} & & \\ & & \ddots & \\ & & & P_d^{-1} \end{pmatrix} (Cy \otimes \chi)$$

$$= \sum_{j=1}^{d} (E_{jj} Cy) \otimes (P_j^{-1} \chi)$$

$$= \left(\left(\sum_{j=1}^{d} \lambda_j^{-1} E_{jj} \right) Cy \right) \otimes \chi$$

$$= \left(\begin{pmatrix} \lambda_1^{-1} & & & \\ & \lambda_2^{-1} & & \\ & & \ddots & \\ & & & \lambda_d^{-1} \end{pmatrix} Cy \right) \otimes \chi.$$

Thus,

$$U(y \otimes \chi) = \alpha(y \otimes \chi)$$

if and only if

$$\begin{pmatrix} \lambda_1^{-1} & & & \\ & \lambda_2^{-1} & & \\ & & \ddots & \\ & & & \lambda_d^{-1} \end{pmatrix} Cy = \alpha y. \qquad \square$$

7.2 Commuting Shunts and Grover Coins

In this section, we study the spectrum of a shunt-decomposition walk where all the shunts commute. A complete characterization follows from Lemma 7.1.2. We then look into the case where each vertex receives the Grover coin, and obtain more explicit formulas for the eigenvalues and eigenvectors of U.

Suppose X has shunt-decomposition

$$A(X) = P_1 + \cdots + P_d,$$

where $P_j P_k = P_k P_j$ for all j and k. Then $P_1, \ldots, P_d$ generate an abelian group Γ, which acts regularly on the vertices of X. Thus, X is isomorphic to a Cayley digraph over Γ with connection set $\{P_1, \ldots, P_d\}$. Since the characters of Γ are eigenvectors for the regular representation of Γ, and distinct characters are orthogonal, by Lemma 7.1.2, they give rise to a basis of eigenvectors for U.

From now on, let Γ be a finite abelian group, and let X be a Cayley digraph over Γ with connection set $\{g_1, \ldots, g_d\}$, denoted

$$X(\Gamma, \{g_1, \ldots, g_d\}).$$

Since Γ is abelian, the images of the connection set under the regular representation of Γ are the shunts $P_1, \ldots, P_d$ in a shunt-decomposition of X. If χ is an character of Γ, then

$$P_j \chi = \chi(g_j) \chi.$$

Define

$$\Lambda_\chi := \begin{pmatrix} \chi(g_1^{-1}) & & & \\ & \chi(g_2^{-1}) & & \\ & & \ddots & \\ & & & \chi(g_d^{-1}) \end{pmatrix}.$$

The following result, as a consequence of Lemma 7.1.2, is again due to Aharonov and colleagues [1].

7.2.1 Theorem. *Let Γ be a finite abelian group. Let X be a Cayley digraph over Γ with connection set $\{g_1,\ldots,g_d\}$. Let U be the transition matrix of a shunt-decomposition C-walk on X. The eigenvalues of U consists of eigenvalues of $\Lambda_\chi C$, where χ ranges over all characters of Γ.* $\qquad\square$

Note that when χ is the trivial character, we have $\Lambda_\chi C = C$. Hence the eigenvalues of the coin are always eigenvalues of U.

Let G be the $d \times d$ Grover coin. Consider a shunt-decomposition Grover walk. We derive explicit formulas for the eigenvalues of U, in terms of the characters. While the following theorem only states what the eigenvalues of U are, the proof also provides a construction for all the eigenvectors.

7.2.2 Theorem. *Let Γ be a finite abelian group. Let X be a Cayley digraph over Γ with connection set $\{g_1,\ldots,g_d\}$. Let U be the transition matrix of a shunt-decomposition Grover walk on X. Let χ be a non-trivial character of Γ. Each eigenvalue α of $\Lambda_\chi G$ is either*

(i) a zero of

$$\frac{1}{\alpha\chi(g_1)+1} + \cdots + \frac{1}{\alpha\chi(g_d)+1} - \frac{d}{2},$$

with multiplicity 1, or

(ii) $-\chi(g_j^{-1})$, with multiplicity one less than the number of k's such that $\chi(g_k) = \chi(g_j)$.

Proof Let y be an eigenvector for $\Lambda_\chi G$ with eigenvalue α. Since

$$G = \frac{2}{d}J - I,$$

we need to solve

$$\frac{2}{d}Jy = (\alpha\Lambda_\chi^{-1} + I)y,$$

that is,

$$\frac{2}{d}\langle \mathbf{1}, y\rangle\mathbf{1} = \begin{pmatrix} (\alpha\chi(g_1)+1)y_1 \\ \vdots \\ (\alpha\chi(g_d)+1)y_d \end{pmatrix}. \tag{7.2.1}$$

Consider two cases.

(i) Suppose $\langle \mathbf{1}, y\rangle \neq 0$. Then the right-hand side in Equation (7.2.1) is a vector with no zero entry. Without loss of generality we may assume $\langle \mathbf{1}, y\rangle = 1$. Thus,

$$\frac{1}{\alpha\chi(g_1)+1} + \cdots + \frac{1}{\alpha\chi(g_d)+1} = \frac{d}{2}, \tag{7.2.2}$$

and each solution α to the above uniquely determines an eigenvector y with $\langle \mathbf{1}, y \rangle = 1$. Therefore, the distinct zeros of Equation (7.2.2) are eigenvalues of $\Lambda_\chi G$. Further, if any of them has multiplicity greater than one, then it must have an eigenvector y such that $\langle y, \mathbf{1} \rangle = 0$, which implies that one of

$$\alpha(\chi(g_1)+1), \ldots, \alpha(\chi(g_d)+1)$$

is equal to zero, a contradiction.

(ii) Suppose $\langle \mathbf{1}, y \rangle = 0$. Since $y \neq 0$, there must exist some j such that

$$\alpha\chi(g_j)+1 = 0,$$

that is, $\alpha = -\chi(g_j^{-1})$. Then for any k such that $\chi(g_j) \neq \chi(g_k)$, we have $y_k = 0$. Hence y is orthogonal to $\mathbf{1}$ if and only if

$$\sum_{k:\chi(g_k)=\chi(g_j)} y_j = 0,$$

from which the multiplicity of α follows. $\qquad\square$

7.3 Uniform Average Vertex Mixing

One topic in the limiting behavior of quantum walks is uniform average vertex mixing. We saw in Section 4.5 that this is guaranteed whenever the average mixing matrix $\widehat{M}$ is flat, or equivalently, when U has simple eigenvalues and flat eigenvectors. However, uniform average vertex mixing does not imply uniform average mixing. The following two results are due to Aharonov and colleagues [1].

7.3.1 Theorem. *Let Γ be a finite abelian group. Let X be a Cayley digraph over Γ with connection set $\{g_1,\ldots,g_d\}$. Let U be the transition matrix of a shunt-decomposition C-walk on X, with spectral decomposition*

$$U = \sum_r e^{i\theta_r} F_r.$$

If U has simple eigenvalues, and for each r, $\langle F_r, \rho_S \rangle = 1$ whenever S is is the set of outgoing arcs of a vertex, then U admits uniform average vertex mixing. $\qquad\square$

7.3.2 Corollary. *Let* Γ *be a finite abelian group. Let* X *be a Cayley digraph over* Γ *with connection set* $\{g_1,\ldots,g_d\}$*. Let* U *be the transition matrix of a shunt-decomposition C-walk on* X*. If* U *has simple eigenvalues, then the quantum walk admits uniform average vertex mixing.*

Proof This follows immediately from the structure of eigenvectors for U, as described in Lemma 7.1.2. $\qquad\square$

Using these results, Aharonov and colleagues [1] showed that on every odd cycle, the shunt-decomposition Hadamard walk admits uniform average mixing. We wish to construct more examples with Grover coins.

Let X be a d-regular Cayley digraph over an abelian group Γ, and let U be the transition matrix of a shunt-decomposition Grover walk on X. The first difficulty we face is that when $d > 2$, the coin G itself contributes -1 to the spectrum of U at least twice. Hence the above corollary no longer applies. Fortunately, simple eigenvalues are not necessary for uniform average vertex mixing to occur; a slightly weaker condition also works.

7.3.3 Theorem. *Let* Γ *be a finite abelian group. Let* X *be a Cayley digraph over* Γ *with connection set* $\{g_1,\ldots,g_d\}$*. Let* U *be the transition matrix of a shunt-decomposition Grover walk on* X*. If the only non-simple eigenvalue of* U *is* -1 *with multiplicity* $d-1$*, then* U *admits uniform average vertex mixing.*

Proof Suppose X has n vertices. Since -1 is an eigenvalue of U with multiplicity $d-1$, by Lemma 7.1.2 and Theorem 7.2.2, the eigenprojection of -1 must be

$$F_{-1} = \left(I - \frac{1}{d}J\right) \otimes \frac{1}{n}J.$$

Let u be any vertex of X, and S the set of outgoing arcs of X. Then

$$F_{-1}\rho_S F_{-1} = \frac{1}{d}\left(\left(I - \frac{1}{d}J\right) \otimes \frac{1}{n}J\right)(I \otimes E_{uu})\left(\left(I - \frac{1}{d}J\right) \otimes \frac{1}{n}J\right)$$

$$= \frac{1}{n^2 d}\left(I - \frac{1}{d}J\right) \otimes (JE_{uu}J)$$

$$= \frac{1}{nd}F_{-1}.$$

On the other hand, since the remaining eigenvalues are all simple, we have

$$F_r\rho_S F_r = \frac{1}{nd}F_r$$

for all $r \neq -1$. Thus, by Theorem 4.3.3,

$$\lim_{K \to \infty} \frac{1}{K} \sum_{k=0}^{K-1} P_{\rho_0,S}(k) = d \sum_{r=1}^{m} \langle F_r \rho_0 F_r, \rho_S \rangle$$

$$= d \sum_r \langle \rho_0, F_r \rho_S F_r \rangle$$

$$= \frac{1}{n} \sum_r \langle \rho_0, F_r \rangle$$

$$= \frac{1}{n}. \qquad \square$$

Combining this with Theorem 7.2.2, we need a Cayley digraph where

$$\chi(g_1), \chi(g_2), \cdots, \chi(g_d)$$

are pairwise distinct, for every non-trivial character χ. This is satisfied when Γ is a cyclic group of prime order.

In the rest of this section, let us assume $X = X(\mathbb{Z}_p, \{g_1, \ldots, g_d\})$ for some prime p. Let

$$\zeta := e^{2\pi i/p}.$$

We wish to characterize circulant digraphs over $\mathbb{Z}_p$ that admit uniform average vertex mixing. To begin, we prove the following easier direction.

7.3.4 Lemma. *Let p be a prime. Let X be a circulant digraph over $\mathbb{Z}_p$ with connection set $\{g_1, \ldots, g_d\}$. Let U be the transition matrix of a shunt-decomposition Grover walk on X. If the connection set of X is fixed by some non-trivial automorphism of $\mathbb{Z}_p$, then there is an initial state for which the average probability distribution is not uniform over the vertices.*

Proof Suppose the connection set is invariant under multiplication by k, for some $k \in \{2, 3, \ldots, p-1\}$. Let χ be the character of $\mathbb{Z}_p$ that sends vertex u to ζ^u, and let ϕ be the character that sends u to ζ^{ku}. Then there is a permutation P such that

$$P \Lambda_\chi P^T = \Lambda_\phi.$$

If y is an eigenvector for $\Lambda_\chi G$ with eigenvalue α, then Py is an eigenvector for $\Lambda_\phi G$ with eigenvalue α. By Lemma 7.1.2, both $y \otimes \chi$ and $Py \otimes \phi$ are eigenvectors for U with eigenvalue α. Choose y such that

$$(y \otimes \chi + Py \otimes \phi)(y \otimes \chi + Py \otimes \phi)^*$$

has trace one; denote this state by ρ_0.

Now let S be the set of outgoing arcs of some vertex u. By Theorem 4.3.3, the average probability on S, given initial state ρ_0, is

$$d \sum_r \langle F_r \rho_0 F_r, \rho_S \rangle = d \langle \rho_0, \rho_S \rangle$$

$$= (y \otimes \chi + Py \otimes \phi)^* (I \otimes E_{uu})(y \otimes \chi + Py \otimes \phi)$$

$$= 2|y|^2 + 2\mathrm{Re}\left(\langle y, Py \rangle e^{2\pi i(k-1)u/p} \right).$$

Since p is a prime, the last line cannot be the same for all vertices u. $\qquad\square$

What about the converse? While we are not able to answer this question in general, we do have the characterization for $d = 2$ and $d = 3$. Some of our techniques may be generalized to circulant digraphs with larger valency.

For $k = 1, 2, \ldots, p - 1$, define

$$f_k(x) := \frac{1}{x - \zeta^{kg_1}} + \cdots + \frac{1}{x - \zeta^{kg_d}} - \frac{d}{2x}.$$

7.3.5 Theorem. *Let p be a prime. Let X be a circulant digraph over $\mathbb{Z}_p$ with connection set $\{g_1, \ldots, g_d\}$. Let U be the transition matrix of a shunt-decomposition Grover walk on X. The eigenvalues of U are 1, -1, and the set of α such that $f_k(-\alpha^{-1}) = 0$ for some $k = 1, 2, \ldots, p - 1$.*

Proof We apply Theorem 7.2.2 to find the eigenvalues of U. Fix a non-trivial character χ of $\mathbb{Z}_p$. Then $\chi(g_j) = \zeta^{kg_j}$ for some $k = 1, 2, \ldots, p - 1$. Since p is a prime, $\chi(g_j)$ is distinct over the connection set, so all eigenvalues of $\Lambda_\chi G$ are of the first type in Theorem 7.2.2. The relation between these eigenvalues and the roots of $f_k(x)$ follows from a simple transformation. $\qquad\square$

7.3.6 Corollary. *Let p be a prime. Let X be a circulant digraph over $\mathbb{Z}_p$ with connection set $\{g_1, \ldots, g_d\}$. Let U be the transition matrix of a shunt-decomposition Grover walk on X. For $k = 1, 2, \ldots, p - 1$, the function $f_k(x)$ has d distinct roots.* $\qquad\square$

By manipulating $f_k(x)$, we find an algebraic relation between the eigenvalues of U and those of X. That is, each eigenvalue of U, other than ± 1, satisfies a polynomial whose coefficients are the eigenvalues of X.

7.3.7 Theorem. *Let p be a prime. Let X be a circulant digraph over $\mathbb{Z}_p$ with connection set $\{g_1, \ldots, g_d\}$. Let U be the transition matrix of a* *shunt-decomposition Grover walk on X. Let $\theta_0, \ldots, \theta_{p-1}$ be the eigenvalues of X. The eigenvalues of U are $1, -1$, and the set of α such that*

$$\frac{d}{2} = \frac{\theta_0^{\sigma_k} + \theta_1^{\sigma_k}(-\alpha) + \cdots + \theta_{p-1}^{\sigma_k}(-\alpha)^{p-1}}{1 - (-\alpha)^p},$$

for some σ_k in the Galois group $\mathrm{Aut}(\mathbb{Q}(\zeta)/\mathbb{Q})$.

Proof Let $\alpha \notin \{-1, 1\}$ be an eigenvalue of U. Let $\beta = -\alpha$. By Theorem 7.3.5, we have $f_k(\beta) = 0$ for some $k = 1, 2, \ldots, p-1$. Thus,

$$\frac{d}{2} = \sum_{j=1}^{d} \frac{1}{1 - \zeta^{kg_j}\beta}$$

$$= \sum_{j=1}^{d} (1 + (\zeta^{kg_j}\beta) + (\zeta^{kg_j}\beta)^2 + \cdots)$$

$$= \sum_{j=1}^{d} \frac{1 + \zeta^{kg_j}\beta + \cdots + (\zeta^{kg_j}\beta)^{p-1}}{1 - \beta^p}$$

$$= \frac{1}{1 - \beta^p} \left(p - 1 + \left(\sum_{j=1}^{d} \zeta^{g_j} \right)^{\sigma_k} \beta + \cdots + \left(\sum_{j=1}^{d} \zeta^{(p-1)g_j} \right)^{\sigma_k} \beta^{p-1} \right).$$

Note that for $\ell = 0, 1, 2, \ldots, p-1$,

$$\sum_{j=1}^{d} \zeta^{\ell g_j}$$

is precisely the ℓth eigenvalue θ_ℓ of X. $\qquad\square$

Both Theorem 7.3.5 and Theorem 7.3.7 give formulas for the eigenvalues of U. Our next goal is to derive a sufficient condition, based on Theorem 7.3.5, for uniform average vertex mixing to happen. Define

$$q_k(x) := (x - \zeta^{kg_1}) \cdots (x - \zeta^{kg_d}).$$

Note that x is a root of $f_k(x)$ if and only if it is a root of

$$h_k(x) := dq_k(x) - 2xq_k'(x).$$

The following is a sufficient condition for uniform average vertex mixing to occur.

7.3.8 Lemma. *Let p be a prime. Let X be a circulant digraph over $\mathbb{Z}_p$ with connection set $\{g_1,\ldots,g_d\}$. Let U be the transition matrix of a shunt-decomposition Grover walk on X. If for any $k = 2, 3, \ldots, p-1$, the polynomials $h_1(x)$ and $h_k(x)$ are coprime over $\mathbb{C}[x]$, then U admits uniform average vertex mixing.*

Proof Recall from Corollary 7.3.6 that each of $h_1(x), \ldots, h_{p-1}(x)$ has d distinct roots. Thus, if $h_1(x), \ldots, h_{p-1}(x)$ are pairwise coprime over $\mathbb{C}[x]$, then the only non-simple eigenvalue of U is -1 with multiplicity $d-1$, and so uniform average vertex mixing occurs. Since the set

$$\{h_1(x), \ldots, h_{p-1}(x)\}$$

is closed under the action of the Galois group $\mathrm{Aut}(\mathbb{Q}(\zeta)/\mathbb{Q})$, it suffices to assume that $h_1(x)$ and $h_k(x)$ are coprime over $\mathbb{C}[x]$, for $k = 2, 3, \ldots, d$. $\qquad\square$

We apply the above criterion to X with $d = 2$. This is not the most exciting quantum walk, as the 2×2 Grover coin is simply a permutation matrix. However, the result gives some hint on the condition we should impose on the connection set.

7.3.9 Theorem. *Let p be a prime and X a 2-regular circulant digraph over $\mathbb{Z}_p$. Let U be the transition matrix of a shunt-decomposition Grover walk on X. Then U admits uniform average vertex mixing if and only if the connection set is not inverse closed, that is, X is not a graph.*

Proof Let $X = X(\mathbb{Z}_p, \{g_1, g_2\})$. Note that

$$2q_k(\dot{x}) - 2xq_k'(x) = 0$$

if and only if

$$x^2 = \zeta^{k(g_1+g_2)}.$$

Hence $f_1(x) = f_k(x)$ if and only if $g_1 + g_2 = p$. $\qquad\square$

7.4 3-Regular Circulants

We generalize our last theorem to 3-regular circulant digraphs on p vertices, for any prime $p \geq 5$. The analysis becomes much more complicated now. To start, we need the following result on cyclotomic integers.

7.4.1 Lemma. *Let $m \in \mathbb{Z}$ and let $p \geq 5$ be a prime. If m divides a cyclotomic integer*

$$\sum_{j=0}^{p-1} a_j \zeta^j,$$

then

$$a_0 \equiv a_1 \equiv \cdots \equiv a_{p-1} \pmod{m}.$$

Proof The expression

$$\sum_{j=0}^{p-1} a_j \zeta^j$$

of an element in $\mathbb{Z}[\zeta]$ is unique up to summing integer multiples of

$$1 + \zeta + \cdots + \zeta^{p-1}. \qquad \square$$

Next, note that when $d = 3$,

$$h_1(x) = 3x^3 - s_1 x^2 - s_2 x + 3s_3,$$

where s_1, s_2, and s_3 are elementary symmetric functions in ζ^{g_1}, ζ^{g_2}, and ζ^{g_3}:

$$s_1 = \sum_{j=1}^{3} \zeta^{g_j}, \quad s_2 = \sum_{1 \leq j < \ell \leq 3} \zeta^{g_j + g_\ell}, \quad s_3 = \zeta^{g_1 + g_2 + g_3}.$$

Similarly, fixing some $k \in \{2, 3, \ldots, p-1\}$, we can write

$$h_k(x) = 3x^3 - t_1 x^2 - t_2 x + 3t_3,$$

where t_1, t_2, and t_3 are elementary symmetric functions in ζ^{kg_1}, ζ^{kg_2}, and ζ^{kg_3}.

The *resultant* of two polynomials is the determinant of their Sylvester matrix. Given two polynomials over an integral domain, their resultant is zero if and only if they have a common root.

7.4.2 Lemma. *Let $p \geq 5$ be a prime. Let g_1, g_2, and g_3 be three distinct elements in $\mathbb{Z}_p$. Let s_1, s_2, and s_3 be the elementary symmetric functions in ζ^{g_1}, ζ^{g_2}, and ζ^{g_3}. For any $k \in \{2, 3, \ldots, p-1\}$, let t_1, t_2, and t_3 be the elementary symmetric functions in ζ^{kg_1}, ζ^{kg_2}, and ζ^{kg_3}. Let*

$$h_1(x) = 3x^3 - s_1 x^2 - s_2 x + 3s_3,$$

and

$$h_k(x) = 3x^3 - t_1 x^2 - t_2 x + 3t_3.$$

If $h_1(x)$ and $h_k(x)$ share a root, then we have the equality

$$s_1 t_2 = s_2 t_1$$

in $\mathbb{Z}_3[\zeta]$.

Proof The resultant of $h_1(x)$ and $h_k(x)$ is an integer multiple of

$$s_3 t_3 (s_1 - t_1)(\overline{s_1 - t_1})(s_1 t_2 - s_2 t_1) + 3\gamma$$

for some $\gamma \in \mathbb{Z}[\zeta]$. If $h_1(x)$ and $h_k(x)$ share a root, then their resultant is zero and so 3 divides

$$(s_1 - t_1)(\overline{s_1 - t_1})(s_1 t_2 - s_2 t_1)$$

in $\mathbb{Z}[\zeta]$. By Lemma 7.4.1, the expression $(s_1 - t_1)(\overline{s_1 - t_1})(s_1 t_2 - s_2 t_1)$ is a polynomial in ζ whose coefficients are congruent to each other modulo 3. Suppose

$$s_1 t_2 - s_2 t_1 = \sum_{j=0}^{p-1} a_j \zeta^j.$$

Let a be the vector of the coefficients, that is,

$$a = \begin{pmatrix} a_0 \\ a_1 \\ \vdots \\ a_{p-1} \end{pmatrix}.$$

We derive conditions that a needs to satisfy.

Let P be the $p \times p$ circulant permutation matrix

$$P := \begin{pmatrix} 0 & 1 & 0 & \cdots & 0 \\ 0 & 0 & 1 & \cdots & 0 \\ \cdots & \cdots & \cdots & \cdots & \cdots \\ 0 & 0 & 0 & \cdots & 1 \\ 1 & 0 & 0 & \cdots & 0 \end{pmatrix}.$$

Define

$$Q = P^{g_1} + P^{g_2} + P^{g_3} - P^{kg_1} - P^{kg_2} - P^{kg_3}.$$

Then

$$(s_1 - t_1)(\overline{s_1 - t_1})(s_1 t_2 - s_2 t_1)$$

is a polynomial in ζ with entries of $Q^T Q a$ as coefficients. Thus, $Q^T Q a$ is a constant vector over $\mathbb{Z}_3$. On the other hand, both the rows and the columns of Q generate the same cyclic code over $\mathbb{Z}_3$ with dimension $p - 1$, whose dual code is generated by $\mathbf{1}$. Therefore,

$$Q^T Q a \equiv 0 \pmod{3}.$$

It follows that $Qa \equiv 0 \pmod{3}$, and so a must be a constant vector over $\mathbb{Z}_3$. Note that there are no more than 18 non-zero entries in a, so for $p \geq 19$, we must have $a \equiv 0 \pmod{3}$. The cases where $p < 19$ can be easily verified by computation. $\qquad\square$

7.4.3 Lemma. *Let $p \geq 5$ be a prime. Let g_1, g_2, and g_3 be three distinct elements in $\mathbb{Z}_p$. Let s_1, s_2, and s_3 be the elementary symmetric functions in ζ^{g_1}, ζ^{g_2}, and ζ^{g_3}. For any $k \in \{2, 3, \ldots, p - 1\}$, let t_1, t_2, and t_3 be the elementary symmetric functions in ζ^{kg_1}, ζ^{kg_2}, and ζ^{kg_3}. Let*

$$h_1(x) = 3x^3 - s_1 x^2 - s_2 x + 3s_3,$$

and

$$h_k(x) = 3x^3 - t_1 x^2 - t_2 x + 3t_3.$$

If $h_1(x)$ and $h_2(x)$ share a root, then the set $\{g_1, g_2, g_3\}$ is fixed by some non-trivial automorphism of $\mathbb{Z}_p$.

Proof The case where $p = 5$ can be easily verified. Let $p \geq 7$. Suppose $h_1(x)$ and $h_2(x)$ share a root. By Lemma 7.4.2,

$$s_1 t_2 - s_2 t_1$$

is a polynomial $\psi(\zeta)$ in ζ whose coefficients are all divisible by 3. For notational ease, let

$$z_j := \zeta^{g_j}.$$

We expand $s_1 t_2$ and $s_2 t_1$:

$$s_1 t_2 = z_1 z_2^k z_3^k + z_1^k z_2 z_3^k + z_1^k z_2^k z_3 \tag{7.4.1}$$
$$+ z_1^{k+1} z_2^k + z_2^{k+1} z_3^k + z_1^k z_3^{k+1} \tag{7.4.2}$$
$$+ z_1^{k+1} z_3^k + z_1^k z_2^{k+1} + z_2^k z_3^{k+1}. \tag{7.4.3}$$
$$s_2 t_1 = z_1 z_2 z_3^k + z_1 z_2^k z_3 + z_1^k z_2 z_3 \tag{7.4.4}$$
$$+ z_1^{k+1} z_2 + z_2^{k+1} z_3 + z_1 z_3^{k+1} \tag{7.4.5}$$
$$+ z_1^{k+1} z_3 + z_1 z_2^{k+1} + z_2 z_3^{k+1}. \tag{7.4.6}$$

Consider two cases.

(i) All coefficients in $\psi(\zeta)$ are zero. Then there is a bijection between the nine terms in Lines (7.4.1), (7.4.2), (7.4.3) and the nine terms in Lines (7.4.4), (7.4.5), (7.4.6). In particular, both sets of terms have the same product. Thus,

$$s_3^{3k+6} = s_3^{6k+3},$$

and so $k = p - 1$. Combining this with $s_1 t_2 = s_2 t_1$, we have $s_1^2 = s_2^2$. Clearly, $s_1 \neq -s_2$ for $p \geq 7$. If $s_1 = s_2$, playing the same product trick shows that $s_3 = 1$. Now,

$$s_1 = \overline{s_2} s_3 = \overline{s_1},$$

which is impossible.

(ii) Some coefficient in $\psi(\zeta)$ is at least 3. Then at least three of the nine terms in Lines (7.4.1), (7.4.2), (7.4.3) are equal to some value β.

1. One of the three terms in Line (7.4.1), say $z_1 z_2^k z_3^k$, is equal to β. Clearly,

$$\beta \notin \{z_1^k z_2 z_3^k, z_1^k z_2^k z_3, z_1^{k+1} z_2^k, z_2^{k+1} z_3^k, z_1^{k+1} z_3^k, z_2^k z_3^{k+1}\}.$$

Hence we must have

$$\beta = z_1 z_2^k z_3^k = z_1^k z_2^{k+1} = z_1^k z_3^{k+1}. \tag{7.4.7}$$

The last equality implies $k = p - 1$, while the second equality implies $z_1^3 = s_3$. Now

$$s_1 t_2 = 3\overline{z_1} + \overline{z_1} z_2 \overline{z_3} + \overline{z_1} \overline{z_2} z_3 + 2\overline{z_2} + 2\overline{z_3}.$$

It is not hard to verify that

$$\overline{z_1} z_2 \overline{z_3}, \overline{z_1} \overline{z_2} z_3, \overline{z_2}, \overline{z_3}$$

are pairwise distinct. Thus, the last four terms on the right-hand side of Equation 7.4.7 cannot survive in $s_1 t_2 - s_2 t_1$, and from the expansion of $s_2 t_1$, we must have

$$\overline{z_1} z_2 \overline{z_3} + \overline{z_1} \overline{z_2} z_3 + 2\overline{z_2} + 2\overline{z_3} = z_1 \overline{z_2} z_3 + z_1 z_2 \overline{z_3} + 2z_2 + 2z_3.$$

Since $p \geq 7$ is a prime, this can only happen when

$$\{\overline{z_1} z_2 \overline{z_3}, \overline{z_1} \overline{z_2} z_3, \overline{z_2}, \overline{z_3}\} = \{z_1 \overline{z_2} z_3, z_1 z_2 \overline{z_3}, z_2, z_3\}.$$

As a result, both sets have the same product, and so $z_1 = s_3$, which contradicts $z_1^3 = s_3$.

2. No term in Line (7.4.1) is equal to β. Suppose without loss of generality that $\beta = z_1^{k+1} z_2^k$. Clearly,

$$\beta \notin \{z_1^{k+1} z_3^k, z_1^k z_2^{k+1}\}.$$

Also, since

$$z_2^k z_3^{k+1} \notin \{z_2^{k+1} z_3^k, z_1^k z_3^{k+1}\},$$

for β to appear at least three times in Line (7.4.2) and Line (7.4.3), it must be that

$$\beta = z_1^{k+1} z_2^k = z_2^{k+1} z_3^k = z_1^k z_3^{k+1}. \tag{7.4.8}$$

It is not hard to verify that the remaining six terms in Line (7.4.1) and Line (7.4.3) are pairwise distinct, given Equation 7.4.8. Thus, they have to vanish in $s_1 t_2 - s_2 t_1$. Meanwhile,

$$\begin{aligned} s_2 t_1 &= z_1^k z_2 z_3 + z_1 z_2^k z_3 + z_1 z_2 z_3^k \\ &\quad + z_1^{k+1} z_2 + z_2^{k+1} z_3 + z_1 z_3^{k+1} \\ &\quad + 3 z_1^{k+1} z_3. \end{aligned}$$

Thus,

$$\begin{aligned} &z_1 z_2^k z_3^k + z_1^k z_2 z_3^k + z_1^k z_2^k z_3 + z_1^{k+1} z_2^k + z_2^{k+1} z_3^k + z_1^k z_3^{k+1} \\ &= z_1^k z_2 z_3 + z_1 z_2^k z + z_1 z_2 z_3^k + z_1^{k+1} z_2 + z_2^{k+1} z_3 + z_1 z_3^{k+1}. \end{aligned}$$

Again, since $p \geq 7$ is a prime, this implies that the products of terms on both sides are equal. Therefore $s_3 = 1$. It follows from Equation (7.4.8) that

$$z_1^{k+2} = z_3^{k-1}, \quad z_2^{k+2} = z_3^{k-1}, \quad z_3^{k+2} = z_2^{k-1}. \tag{7.4.9}$$

Since $k - 1$ is coprime to p, there exists an integer ℓ such that

$$(k - 1)\ell \equiv 1 \pmod{p}.$$

Therefore, Equation 7.4.9 shows that the connection set is invariant under multiplication by ℓ in $\mathbb{Z}_p$. $\quad\square$

We are now ready to characterize 3-regular circulant digraphs on a prime number of vertices that admit uniform average vertex mixing.

7.4.4 Theorem. *Let p be a prime. Let X be a 3-regular circulant digraph over $\mathbb{Z}_p$. Then the shunt-decomposition Grover walk on X admits uniform average vertex mixing if and only if its connection set has a trivial stabilizer in $\mathrm{Aut}(\mathbb{Z}_p)$.*

Proof Let X be a 3-regular circulant digraph on p vertices. Let U be the transition matrix of the shunt-decomposition Grover walk on X. If the connection set is not fixed by any non-trivial automorphism of $\mathbb{Z}_p$, then by Lemma 7.4.3,

$$\gcd(h_1(x), h_k(x)) = 1$$

for all $k = 2, 3, \ldots, p - 1$. On the other hand, since

$$\{h_1(x), \ldots, h_{p-1}(x)\}$$

is closed under the action of $\mathrm{Aut}(\mathbb{Q}[\zeta]/\mathbb{Q})$, any rational root x_0 of one polynomial must be a common root of the remaining $p - 2$ polynomials. Therefore none of $h_1(x), \ldots, h_{p-1}(x)$ has 1 or -1 as a root. The result then follows from Theorem 7.3.3. $\qquad\qquad\square$

A special case of 3-regular circulant digraphs is $X(\mathbb{Z}_n, \{1, -1, a\})$, that is, a cycle plus a shunt. Quantum walks on such digraphs are known as "lively quantum walks," and have been studied in [62].

7.5 Unitary Covers

When a particular model does not exhibit the desired property, our quantum walker seeks alternatives by changing the coins, the shunt-decompositions, or even the operator itself. One thing she has not tried, though, is to enlarge the state space she lives in.

We now introduce quantum walks on unitary covers of digraphs; they can be seen as quantum walks on the base digraphs with enlarged state spaces. Our model generalizes the shunt-decomposition walks [1], as well as the Möbius quantum walks [54].

For a digraph X, an r-fold unitary arc function is a map ϕ from the arcs of X to the unitary group of degree r. Given a shunt-decomposition of X, we define what it means for ϕ to be "compatible with" the shunt-decomposition; such a unitary arc function is called a shunt function. Finally, for a digraph X with shunt function ϕ, we construct a quantum walk on X^ϕ, and study the spectral decomposition of its transition matrix. Numerical experiments show that such a walk allows uniform average vertex mixing to occur on X, even if it is impossible on the usual shunt-decomposition walk.

We have seen how covers give rise to interesting walks in Chapter 6. There are two parts in the definition of a cover that we can generalize: the underlying graph, and the arc function.

Let X be a connected digraph. A *unitary arc function of index* r of X is a map ϕ from the arcs of X to $U(r)$, the unitary group of degree r, such that

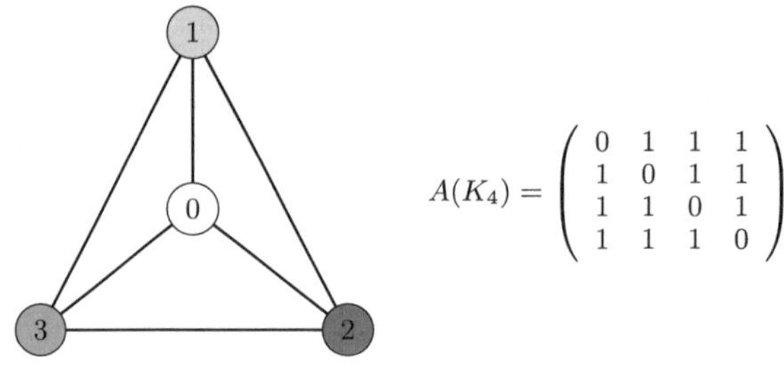

$$A(K_4) = \begin{pmatrix} 0 & 1 & 1 & 1 \\ 1 & 0 & 1 & 1 \\ 1 & 1 & 0 & 1 \\ 1 & 1 & 1 & 0 \end{pmatrix}$$

Figure 7.1 K_4 and its adjacency matrix

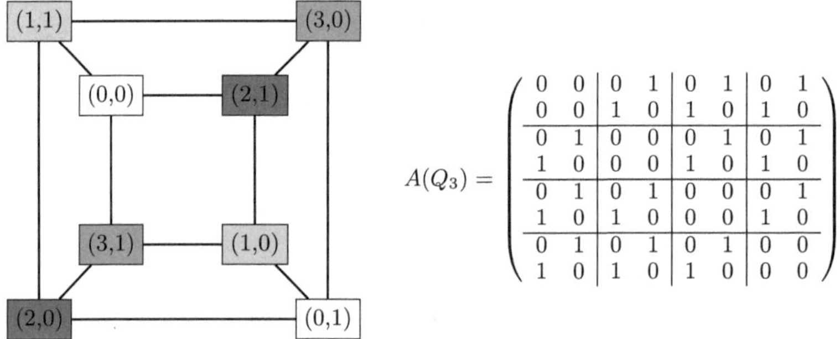

$$A(Q_3) = \left(\begin{array}{cc|cc|cc|cc} 0 & 0 & 0 & 1 & 0 & 1 & 0 & 1 \\ 0 & 0 & 1 & 0 & 1 & 0 & 1 & 0 \\ \hline 0 & 1 & 0 & 0 & 0 & 1 & 0 & 1 \\ 1 & 0 & 0 & 0 & 1 & 0 & 1 & 0 \\ \hline 0 & 1 & 0 & 1 & 0 & 0 & 0 & 1 \\ 1 & 0 & 1 & 0 & 0 & 0 & 1 & 0 \\ \hline 0 & 1 & 0 & 1 & 0 & 1 & 0 & 0 \\ 1 & 0 & 1 & 0 & 1 & 0 & 0 & 0 \end{array} \right)$$

Figure 7.2 Q_3 and its adjacency matrix

$\phi(u, v) = \phi(v, u)^{-1}$. Let $A(X)^\phi$ be the matrix obtained from $A(X)$ by replacing A_{uv} with $\phi(u, v)$ if (u, v) is an arc of X, and with an $r \times r$ block of zeros otherwise. The weighted digraph X^ϕ underlying $A(X)^\phi$ is called a *unitary r-fold cover*.

When the image of ϕ consists of only permutation matrices, we omit the word "unitary" and call ϕ an *r-fold cover*. If in addition X is undirected, then we are back to the special case in Section 5.1. Recall that X^ϕ can be built as follows: replace each vertex u of X by its fiber,

$$\{(u, j) : i = 0, 1, \ldots, r - 1\},$$

and join (u, j) to (v, k) whenever $\phi(u, v)(j) = k$. Alternatively, a digraph Y covers X if there is a homomorphism ψ from Y to X, such that for any vertex y of Y and $x = \psi(y)$, the homomorphism restricted to the outgoing arcs of y in Y is a bijection onto the outgoing arcs of x in X.

As shown in Figure 7.1 and Figure 7.2, the hypercube Q_3 is a double cover of the complete graph K_4, with covering map ψ given by the vertex colouring, The arc function ϕ sends every arc of K_4 to $(1, 2) \in \text{Sym}(2)$. We attach their adjacency matrices to illustrate the construction from ϕ.

Most discussion of covers focuses on voltage graphs; see, for example, [38]. The *orthogonal covers*, for which the image of ϕ consists of only orthogonal matrices, have also been studied in [33]. In this chapter, we will consider unitary covers that respect shunt-decompositions, for they may preserve nice properties that the underlying digraphs admit.

7.6 Shunt Functions

Let X be a d-regular digraph on n vertices, with shunt-decomposition

$$A(X) = P_1 + \cdots + P_d.$$

We are interested in unitary arc functions ϕ that are "compatible with" the shunt-decomposition, that is,

(i) for each arc (u, v), the value $\phi(u, v)$ depends only on the shunt (u, v) belongs to;
(ii) whenever P_j and P_j^T both appear as shunts, we have

$$\phi(P_j)^{-1} = \phi(P_j^T).$$

A unitary arc function ϕ satisfying (i) and (ii) is called a *shunt function*.

Given a shunt function ϕ, we define a quantum walk on X^ϕ as follows. Pick a $d \times d$ unitary coin C. The shift matrix S is a $dnr \times dnr$ block diagonal matrix:

$$S = \begin{pmatrix} P_1 \otimes \phi(P_1) & & & \\ & P_2 \otimes \phi(P_2) & & \\ & & \ddots & \\ & & & P_d \otimes \phi(P_d) \end{pmatrix},$$

and the coin matrix is a $dnr \times dnr$ unitary matrix of the form

$$C \otimes I_n \otimes I_r.$$

Our new quantum walk on X^ϕ, called the *shunt-function walk*, is then determined by the transition matrix

$$U := S(C \otimes I_n \otimes I_r).$$

We explain the connection between this walk and the ones in [1] and [54]. If $r = 1$ and ϕ is the identity map, then $X^\phi = X$ and the shift matrix is simply

$$S = \begin{pmatrix} P_1 & & & \\ & P_2 & & \\ & & \ddots & \\ & & & P_d \end{pmatrix}.$$

Thus, U coincides with the transition matrix of a shunt-decomposition walk on X. On the other hand, if X is the n-cycle C_n, there is a shunt-decomposition

$$A(C_n) = P + P^{-1}$$

where P is cyclic of order n. Suppose in addition that

$$\phi(P) = \begin{pmatrix} \cos\left(\frac{\theta}{2}\right) & i\sin\left(\frac{\theta}{2}\right) \\ i\sin\left(\frac{\theta}{2}\right) & \cos\left(\frac{\theta}{2}\right) \end{pmatrix}$$

and

$$\phi(P^{-1}) = \begin{pmatrix} \cos\left(-\frac{\theta}{2}\right) & i\sin\left(-\frac{\theta}{2}\right) \\ i\sin\left(-\frac{\theta}{2}\right) & \cos\left(-\frac{\theta}{2}\right) \end{pmatrix};$$

then our walk is precisely the Möbius walk defined in [54].

7.7 Spectral Decomposition

To simplify our analysis, we assume all shunts commute, and

$$\phi(P_j)\phi(P_k) = \phi(P_k)\phi(P_j).$$

The following lemma shows how to obtain the spectral decomposition of a shunt-function walk.

7.7.1 Lemma. *Let X be a d-regular graph on n vertices with shunt-decomposition*

$$A(X) = P_1 + \cdots + P_d.$$

Let ϕ be a shunt function of index r. Let

$$U = S(C \otimes I_n \otimes I_r)$$

be the transition matrix of a shunt-function walk on X^ϕ. Let y be a common eigenvector of the shunts, with

$$P_j y = \lambda_j y.$$

Let z be a common eigenvector of $\phi(P_1), \phi(P_2), \ldots, \phi(P_d)$, with

$$\phi(P_j)z = \mu_j z.$$

Let x be a vector of length d. Then $x \otimes y \otimes z$ is an eigenvector of U for the eigenvalue α if and only if x is an eigenvector of

$$\begin{pmatrix} \lambda_1\mu_1 & & \\ & \ddots & \\ & & \lambda_d\mu_d \end{pmatrix} C$$

for the eigenvalue α.

Proof Rewrite

$$S = \sum_j E_{jj} \otimes P_j \otimes \phi(P_j).$$

Then

$$U = \sum_j E_{jj}C \otimes P_j \otimes \phi(P_j).$$

Let

$$D := \begin{pmatrix} \lambda_1\mu_1 & & \\ & \ddots & \\ & & \lambda_d\mu_d \end{pmatrix}.$$

We have

$$U(x \otimes y \otimes z) = \sum_j E_{jj}Cx \otimes P_jy \otimes \phi(P_j)z$$

$$= \left(\sum_j \lambda_j\mu_jE_{jj} \right) Cx \otimes y \otimes z$$

$$= DCx \otimes y \otimes z.$$

Thus, $x \otimes y \otimes z$ is an eigenvector of U for the eigenvalue α if and only if x is an eigenvector of DC for the eigenvalue α. $\qquad\square$

Using an argument similar to the proof of Theorem 3.6 in [1], we see that simple eigenvalues of U guarantees uniform average vertex mixing.

7.7.2 Lemma. *Let X be a Cayley digraph over an abelian group, with shunt-decomposition*

$$A(X) = P_1 + \cdots + P_d.$$

Let ϕ be a shunt function of index r, such that for all j and k, we have $\phi(P_j) = \phi(P_k)$. Let

$$U = S(C \otimes I_n \otimes I_r)$$

be the transition matrix of a shunt-function walk on X^ϕ. If U has simple eigenvalues, then U admits uniform average vertex mixing. $\qquad\square$

Notes

Compared to the arc-reversal walk, the shunt-decomposition walk has better mixing properties, but less search capacity.

A 3-regular circulant digraph over $\mathbb{Z}_p$ admits uniform average vertex mixing if and only if its connection set has a trivial stabilizer in $\mathrm{Aut}(\mathbb{Z}_p)$. This provides the first infinite family of digraphs, other than cycles, that admit uniform average mixing. We believe a similar characterization works when the degree is greater than three.

(i) **Conjecture.** Let p be a prime. A circulant digraph on p vertices admits uniform average vertex mixing if and only if its connection set has a trivial stabilizer in $\mathrm{Aut}(\mathbb{Z}_p)$.

(ii) Find an example with arc-state transfer at time k, with $U^k \neq \pm I$.

(iii) Does uniform average vertex mixing occur on irregular graphs?

8

1-Dimensional Walks

8.1 Infinite Paths

In this chapter, we briefly discuss another approach to shunt-decomposition walks on infinite graphs, based on the paper by Ambainis, Bach, Nayak, Vishwanath, and Watrous [3].

While a lot has been done since Aharonov and colleagues [1] introduced the shunt-decomposition model, most work focused on presenting numerical results. The first paper with exact analysis was due to Ambainis and colleagues [3], who studied the limiting behavior of a shunt-decomposition Hadamard walk on the infinite path P_∞.

As usual, the quantum walker moves on the arcs of P_∞. The state space can be identified as $\mathbb{C}^{\mathbb{Z}} \otimes \mathbb{C}^2$ (more formally, $\ell_2(\mathbb{C}^{\mathbb{Z}}) \otimes \mathbb{C}^2$)). Recall that the Hadamard coin is the 2×2 Fourier coin:

$$H := \frac{1}{\sqrt{2}} \begin{pmatrix} 1 & 1 \\ 1 & -1 \end{pmatrix}.$$

Let S be the linear operator such that

$$S(e_u \otimes e_1) = e_{u+1} \otimes e_1$$

and

$$S(e_u \otimes e_2) = e_{u-1} \otimes e_2.$$

Then the transition operator of the shunt-decomposition walk with Hadamard coins is

$$U = S(H \otimes I).$$

We will refer to this walk as the Hadamard walk.

Given initial state $e_0 \otimes e_1$, let $\Psi(u, k)$ be the coin state on vertex u at time k. Ambainis and colleagues [3] derived a recurrence relation for $\Psi(u, k)$:

$$\Psi(u, k+1) = \frac{1}{\sqrt{2}} \begin{pmatrix} 0 & 0 \\ 1 & 1 \end{pmatrix} \Psi(u - 1, k) + \frac{1}{\sqrt{2}} \begin{pmatrix} -1 & 1 \\ 0 & 0 \end{pmatrix} \Psi(u + 1, k),$$

121

with initial conditions

$$\Psi(0,0) = \begin{pmatrix} 0 \\ 1 \end{pmatrix}$$

and

$$\Psi(u,0) = \begin{pmatrix} 0 \\ 0 \end{pmatrix},$$

for all $u \neq 0$. Using this recurrence, they proved several properties of the probability distribution, all strikingly different from the classical random walk on P_∞. For example, after k steps, the probability distribution of this Hadamard walk is nearly uniform over the vertices between $-k/\sqrt{2}$ and $k/\sqrt{2}$, while a classical random walker tends to stay at distance $O(\sqrt{k})$ from the origin with high probability. In the presence of absorbing boundaries, the exit probabilities are also in sharp contrast to those of the classical random walk. With one absorbing boundary at vertex 0, the probability that the walker exits to the left is $2/\pi$, and with an additional absorbing boundary at vertex u, this probability increases, and approaches $1/\sqrt{2}$ as u goes to infinity. Both probabilities in the classical random walk are 1.

8.2 Coupling Walks

Our goal to is analyse the Hadamard walk on the infinite path. To do this we will view the infinite path as the limit of the finite cycles C_n as $n \to \infty$; but, before considering walks on cycles, we introduce a more general class of walks.

Let U_0 and U_1 be two $n \times n$ unitary matrices. Observe that $\mathbb{C}^n \otimes \mathbb{C}^2$ is the direct sum of the subspaces

$$\mathbb{C}^n \otimes e_0, \quad \mathbb{C}^n \otimes e_1;$$

if we view U_0 as acting on the first of these subspaces and U_1 as acting on the second, then the direct sum of operators

$$\begin{pmatrix} U_0 & 0 \\ 0 & U_1 \end{pmatrix}$$

is a unitary operator on $\mathbb{C}^n \otimes \mathbb{C}^2$ that fixes the two summands above. Hence it determines a discrete walk on $\mathbb{C}^n \otimes \mathbb{C}^2$, which we may view as the direct sum of the walks given by U_0 and U_1.

Now set

$$H = \frac{1}{\sqrt{2}} \begin{pmatrix} 1 & 1 \\ 1 & -1 \end{pmatrix}$$

and define

$$U := \begin{pmatrix} U_0 & 0 \\ 0 & U_1 \end{pmatrix} H = \frac{1}{\sqrt{2}} \begin{pmatrix} U_0 & U_0 \\ U_1 & -U_1 \end{pmatrix}.$$

We say that U is a *coupling* of the walks determined by U_0 and U_1. (Note that we could replace H by any 2×2 unitary matrix.)

If we assume that U_0 and U_1 commute, we can determine the eigenvalues and eigenvectors of U in terms of the eigenvalues and eigenvectors of U_0 and U_1. The first step is to note that, since U_0 and U_1 are normal and commute, there is an orthonormal basis $z_1, \ldots, z_n$ of $\mathbb{C}^n$ consisting of common eigenvectors for U_0 and U_1.

8.2.1 Lemma. *Let U_0 and U_1 be commuting $n \times n$ unitary matrices, let z be a common eigenvector for these matrices, and assume*

$$U_0 z = \theta_0 z, \quad U_1 z = \theta_1 z.$$

Then the two roots of

$$t^2 - \frac{1}{\sqrt{2}}(\theta_1 - \theta_0)t - \theta_0 \theta_1$$

are eigenvalues of U.

Proof Since the entries in

$$tI - U = \begin{pmatrix} tI - \frac{1}{\sqrt{2}} U_0 & -\frac{1}{\sqrt{2}} U_0 \\ -\frac{1}{\sqrt{2}} U_1 & tI + \frac{1}{\sqrt{2}} U_1 \end{pmatrix}$$

commute,

$$\det \begin{pmatrix} tI - \frac{1}{\sqrt{2}} U_0 & -\frac{1}{\sqrt{2}} U_0 \\ -\frac{1}{\sqrt{2}} U_1 & tI + \frac{1}{\sqrt{2}} U_1 \end{pmatrix}$$
$$= \det \left(\left(tI - \frac{1}{\sqrt{2}} U_0 \right) \left(tI + \frac{1}{\sqrt{2}} U_1 \right) - \frac{1}{2} U_0 U_1 \right).$$

Suppose z is a common eigenvector for U_0 and U_1 and the associated eigenvalues are θ_0 and θ_1. Then the eigenvalues of

$$\left(tI - \frac{1}{\sqrt{2}} U_0 \right) \left(tI + \frac{1}{\sqrt{2}} U_1 \right) - \frac{1}{2} U_0 U_1$$

are the roots of

$$t^2 - \frac{1}{\sqrt{2}}(\theta_1 - \theta_0)t - \theta_0\theta_1. \qquad \square$$

8.3 Spectral Decomposition

In the only cases of interest to us, $U_1 = U_0^{-1}$ and hence if θ is an eigenvalue of U_0, then θ^{-1} is the corresponding eigenvalue of θ_1. Under this assumption, we compute the spectral decomposition of U.

8.3.1 Lemma. *Assume* $U_1 = U_0^{-1}$ *and let* θ *be an eigenvalue of* U_0. *If* $\theta = e^{i\varphi}$, *then the two associated eigenvalues of* U *are*

$$\frac{1}{\sqrt{2}}(-i\sin\varphi \pm \sqrt{\cos^2\varphi + 1}).$$

Proof The roots of

$$t^2 - \frac{1}{\sqrt{2}}(\theta^{-1} - \theta) - 1$$

are

$$\frac{1}{2}\left(\frac{\theta^{-1} - \theta}{\sqrt{2}} \pm \sqrt{\frac{(\theta - \theta^{-1})^2}{2} + 4}\right)$$

which we can rewrite as

$$\frac{1}{\sqrt{2}}\left(-\frac{\theta - \theta^{-1}}{2} \pm \sqrt{\frac{(\theta + \theta^{-1})^2}{2} + 1}\right).$$

Next we determine the eigenvectors of U.

8.3.2 Lemma. *Assume*

$$U = \frac{1}{\sqrt{2}}\begin{pmatrix} U_0 & U_0 \\ U_0^{-1} & -U_0^{-1} \end{pmatrix}$$

and let z *be an eigenvector for* U_0 *with norm one and eigenvalue* θ. *If*

$$\lambda := \frac{1}{2\theta}(-\theta - \theta^{-1} \pm \sqrt{\theta + \theta^{-1} + 6}),$$

then

$$\begin{pmatrix} z \\ \lambda z \end{pmatrix}$$

is an eigenvector for U with eigenvalue

$$\frac{1}{\sqrt{2}}\left(-\frac{\theta - \theta^{-1}}{2} \pm \sqrt{\frac{(\theta + \theta^{-1})^2}{2} + 1}\right).$$

Proof We have

$$\begin{pmatrix} U_0 & U_0 \\ U_0^{-1} & -U_0^{-1} \end{pmatrix}\begin{pmatrix} z \\ \lambda z \end{pmatrix} = \begin{pmatrix} \theta(1 + \lambda)z \\ \theta^{-1}(1 - \lambda)z \end{pmatrix},$$

and therefore

$$\begin{pmatrix} z \\ \lambda z \end{pmatrix}$$

is an eigenvector if

$$\theta^{-2}\frac{(1 - \lambda)}{(1 + \lambda)} = \lambda;$$

solving this for λ yields the stated formula. $\qquad\square$

The projection onto the subspace spanned by the the eigenvector

$$\begin{pmatrix} z \\ \lambda z \end{pmatrix}$$

is

$$\frac{1}{1 + \lambda\bar{\lambda}}\begin{pmatrix} 1 & \bar{\lambda} \\ \lambda & \lambda\bar{\lambda} \end{pmatrix} \otimes zz^*.$$

For later use, we compute $\lambda\bar{\lambda}$ explicitly. Note that $\bar{\theta} = \theta^{-1}$ and, as

$$\lambda = \frac{1}{2\theta}\left(-\theta - \theta^{-1} \pm \sqrt{\theta + \theta^{-1} + 6}\right),$$

we have

$$\bar{\lambda} = \frac{\theta}{2}\left(-\theta - \theta^{-1} \pm \sqrt{\theta + \theta^{-1} + 6}\right).$$

Therefore

$$\lambda\bar{\lambda} = \frac{1}{4}\left((\theta + \theta^{-1})^2 \mp 2(\theta + \theta^{-1})\sqrt{\theta + \theta^{-1} + 6} + \theta + \theta^{-1} + 6\right).$$

8.4 Computing Powers

We derive an expression for U^k, in terms of generating functions. View U as a 2×2 block matrix and let $\Phi_{1,1}(k)$ denote $(U^k)_{1,1}$; define $\Phi_{2,1}$ and so on similarly. Then

$$\begin{pmatrix} \Phi_{1,1}(k+1) \\ \Phi_{2,1}(k+1) \end{pmatrix} = U \begin{pmatrix} \Phi_{1,1}(k) \\ \Phi_{2,1}(k) \end{pmatrix}. \tag{8.4.1}$$

Note that $\Phi_{1,1}(0) = I$ and $\Phi_{2,1}(0) = 0$. Define generating functions

$$\Psi_{1,1}(t) = \sum_{k \geq 0} t^k \Phi_{1,1}(k), \quad \Psi_{2,1}(t) = \sum_{k \geq 0} t^k \Phi_{2,1}(k).$$

If we multiply both sides of (8.4.1) by t^{k+1} and sum over k, we get

$$\begin{pmatrix} \Phi_{1,1}(t) \\ \Phi_{2,1}(t) \end{pmatrix} - \begin{pmatrix} I \\ 0 \end{pmatrix} = U \begin{pmatrix} \Phi_{1,1}(t) \\ \Phi_{2,1}(t) \end{pmatrix}$$

and therefore

$$\begin{pmatrix} \Phi_{1,1}(t) \\ \Phi_{2,1}(t) \end{pmatrix} = \left(I - tU \right)^{-1} \begin{pmatrix} I \\ 0 \end{pmatrix}.$$

We now compute $(I - tU)^{-1}$. We have

$$\begin{pmatrix} I - \frac{1}{\sqrt{2}}tU_0 & -\frac{1}{\sqrt{2}}tU_0 \\ -\frac{1}{\sqrt{2}}tU_1 & I + \frac{1}{\sqrt{2}}tU_1 \end{pmatrix} \begin{pmatrix} I + \frac{1}{\sqrt{2}}tU_1 & \frac{1}{\sqrt{2}}tU_0 \\ \frac{1}{\sqrt{2}}tU_1 & I - \frac{1}{\sqrt{2}}tU_0 \end{pmatrix}$$
$$= \begin{pmatrix} I - \frac{1}{\sqrt{2}}t(U_0 - U_1) - t^2 U_0 U_1 & 0 \\ 0 & I - \frac{1}{\sqrt{2}}t(U_0 - U_1) - t^2 U_0 U_1 \end{pmatrix},$$

from which it follows that

$$\Psi_{1,1}(t) = \left(I - t(U_0 - U_1) - 2t^2 U_0 U_1 \right)^{-1} \left(I + \frac{1}{\sqrt{2}}tU_1 \right)$$

and

$$\Psi_{2,1}(t) = \left(I - t(U_0 - U_1) - 2t^2 U_0 U_1 \right)^{-1} \frac{1}{\sqrt{2}}tU_1.$$

If we make the additional assumption that $U_1 = U_0^{-1}$, then

$$\Psi_{1,1}(t) = \left((1 - t^2)I - \frac{1}{\sqrt{2}}t(U_0 - U_0^{-1}) \right)^{-1} (I + tU_1).$$

If U_0 has the spectral decomposition

$$U_0 = \sum_\theta \theta E_\theta,$$

then $\Psi_{1,1}(t)$ has the spectral decomposition

$$\sum_\theta \frac{1 + t\theta^{-1}/\sqrt{2}}{1 - (\theta - \theta^{-1})t/\sqrt{2} - t^2} E_\theta. \tag{8.4.2}$$

8.5 Extracting Coefficients

We derive an explicit expression for the coefficient of t^k in $\Psi_{1,1}(t)$.

If $F(t)$ is a power series in t over a ring (e.g., the ring of $n \times n$ matrices) we use $[t^k, F(t)]$ to denote the coefficient of t^k in $F(t)$.

8.5.1 Theorem. *If α is a root of*

$$t^2 - \frac{1}{\sqrt{2}}(\theta^{-1} - \theta)t - 1,$$

then

$$\left[t^k, \Psi_{1,1}(t) \right] = \sum_\theta \left(\frac{\alpha^{-1} + \theta^{-1}/\sqrt{2}}{\alpha^{-1} + \alpha} \alpha^{-k} + \frac{\alpha - \theta^{-1}/\sqrt{2}}{\alpha^{-1} + \alpha}(-\alpha)^k \right) E_\theta.$$

Proof If α denotes a root corresponding to the eigenvalue θ of U_0, then

$$1 + \frac{1}{\sqrt{2}}(\theta^{-1} - \theta)t - t^2 = (1 - \alpha^{-1}t)(1 + \alpha t).$$

Thus,

$$\frac{1}{1 + (\theta^{-1} - \theta)t/\sqrt{2} - t^2} = \frac{1}{(1 - \alpha^{-1}t)(1 + \alpha t)}$$

$$= \frac{1}{1 + \alpha^2} \frac{1}{1 - \alpha^{-1}t} + \frac{\alpha^2}{1 + \alpha^2} \frac{1}{1 + \alpha t}.$$

Therefore,

$$\left[t^k, \frac{1}{1 + (\theta^{-1} - \theta)t/\sqrt{2} - t^2} \right] = \frac{1}{1 + \alpha^2}(\alpha^{-1})^k + \frac{\alpha^2}{1 + \alpha^2}(-\alpha)^k$$

and

$$\left[t^k, \frac{\theta^{-1}t/\sqrt{2}}{1+(\theta^{-1}-\theta)t/\sqrt{2}-t^2} \right] = \frac{\theta^{-1}/\sqrt{2}}{1+\alpha^2}(\alpha^{-1})^{k-1} + \frac{\theta^{-1}\alpha^2/\sqrt{2}}{1+\alpha^2}(-\alpha)^{k-1}$$

from which the result follows. $\square$

We can compute $[t^k, \Psi_{2,1}(t)]$ similarly. The following lemma gives $[t^k, \Psi_{1,2}(t)]$ and $[t^k, \Psi_{2,2}(t)]$.

8.5.2 Lemma. *We have*

$$\phi_{1,2}(k) = \overline{\phi_{2,1}(k)}, \quad \phi_{2,2}(k) = -\overline{\phi_{1,1}(k)}.$$

Proof View U as a 2×2 matrix over the ring of complex polynomials in U_0 and U_1 (which commute) and note that $\det(U) = -1$. Any 2×2 unitary matrix over $\mathbb{C}$ with determinant 1 has the form

$$\begin{pmatrix} a & b \\ -\bar{b} & \bar{a} \end{pmatrix}.$$

Multiplication (on the left) by the matrix

$$\begin{pmatrix} 1 & 0 \\ 0 & -1 \end{pmatrix}$$

maps 2×2 unitaries with determinant 1 to unitaries with determinant -1. Hence a 2×2 unitary with determinant -1 has the form

$$\begin{pmatrix} a & b \\ \bar{b} & -\bar{a} \end{pmatrix}$$

and therefore

$$U^k = \begin{pmatrix} \Psi_{1,1}(k) & \overline{\Psi_{2,1}(k)} \\ \Psi_{2,1}(k) & -\overline{\Psi_{1,1}(k)} \end{pmatrix}. \qquad \square$$

Now specialize to the case where U_0 is P, the permutation matrix representing an n-cycle. If we set $\lambda = e^{2\pi i/n}$, the eigenvalues of U_0 are the distinct powers λ^j of λ, and if E_j is the idempotent corresponding to the eigenvalue λ^j, we have

$$(E_j)_{\ell,m} = \frac{1}{2n}\lambda^{j(\ell-m)}.$$

Accordingly, if α_j is the root with positive real part of

$$t^2 - \frac{1}{\sqrt{2}}(\lambda^{-j} - \lambda^j)t - 1,$$

then

$$\left[t^k, \Psi_{1,1}(t) \right]_{\ell,m}$$

$$= \frac{1}{2n} \sum_j \left(\frac{\alpha_j^{-1} + \lambda^{-j}/\sqrt{2}}{\alpha_j^{-1} + \alpha_j} \alpha_j^{-k} + \frac{\alpha_j - \lambda^{-j}/\sqrt{2}}{\alpha_j^{-1} + \alpha_j} (-\alpha_j)^k \right) \lambda^{j(\ell-m)}.$$

If we allow n to tend to infinity and let α denote the root with positive real part of

$$t^2 - \frac{1}{\sqrt{2}} (\eta^{-1} - \eta)t - 1,$$

the above sum converges to an integral over the unit circle in the complex plane

$$\frac{1}{2\pi} \oint \left(\frac{\alpha^{-1} + \eta^{-1}/\sqrt{2}}{\alpha^{-1} + \alpha} \alpha^{-k} + \frac{\alpha - \eta^{-1}/\sqrt{2}}{\alpha^{-1} + \alpha} (-\alpha)^k \right) \eta^{\ell-m} \, d\eta. \qquad (8.5.1)$$

By Lemma 8.2.1, α is an eigenvalue of U, so $|\alpha| = 1$. We may assume $\alpha = e^{i\beta}$ and $\eta = e^{i\theta}$. Then

$$\alpha - \alpha^{-1} = \frac{1}{\sqrt{2}} (\eta^{-1} - \eta).$$

Hence,

$$\sin \beta = -\frac{1}{\sqrt{2}} \sin \theta.$$

Now suppose α has positive real part. We have

$$\cos \beta = \frac{1}{\sqrt{2}} \sqrt{1 + \cos^2 \theta}.$$

So

$$\alpha^{-1} + \frac{1}{\sqrt{2}} \eta^{-1} = \frac{1}{\sqrt{2}} \left(\sqrt{1 + \cos^2 \theta} + \cos \theta \right)$$

and

$$\frac{\alpha^{-1} + \eta^{-1}/\sqrt{2}}{\alpha^{-1} + \alpha} = \frac{1}{2} \left(1 + \frac{\cos \theta}{\sqrt{1 + \cos^2 \theta}} \right).$$

It follows that

$$\oint \left(\frac{\alpha^{-1} + \eta^{-1}/\sqrt{2}}{\alpha^{-1} + \alpha} \alpha^{-k} \right) \eta^{\ell-m}$$

$$d\eta = \frac{i}{2} \int_0^{2\pi} \left(1 + \frac{\cos \theta}{\sqrt{1 + \cos^2 \theta}} \right) e^{i((\ell-m+1)\theta - k\beta)} d\theta$$

and

$$\oint \left(\frac{\alpha^{-1} - \eta^{-1}/\sqrt{2}}{\alpha^{-1} + \alpha} \alpha^{-k} \right) \eta^{\ell-m} d\eta$$

$$= \frac{i}{2} \int_0^{2\pi} \left(1 - \frac{\cos\theta}{\sqrt{1+\cos^2\theta}} \right) e^{i((\ell-m+1)\theta - k\beta)} d\theta$$

$$= \frac{i}{2} \int_{-\pi}^{\pi} \left(1 - \frac{\cos(\theta+\pi)}{\sqrt{1+\cos^2(\theta+\pi)}} \right) e^{i((\ell-m+1)(\theta+\pi)-k(\beta+\pi))} d\theta$$

$$= (-1)^{\ell-m+1-k} \frac{i}{2} \int_{-\pi}^{\pi} \left(1 + \frac{\cos(\theta)}{\sqrt{1+\cos^2(\theta)}} \right) e^{i((\ell-m+1)\theta - k\beta)} d\theta.$$

Hence the integral in 8.5.1 reduces to

$$\frac{i}{2\pi} (1 + (-1)^{\ell-m+1-k}) \int_{-\pi}^{\pi} \left(1 + \frac{\cos\theta}{\sqrt{1+\cos^2\theta}} \right) e^{i((\ell-m+1)\theta - k\beta)} d\theta.$$

This agrees with the second formula in Lemma 7 of [3, Lemma 7]; the readers are invited to derive the first formula, using a similar argument.

Glossary

E_{ij} 01-matrix whose ij-entry is 1. 35

J all-ones matrix. 6

$M \succeq N$ matrix $M - N$ is positive semidefinite. 42

U transition matrix. xi

X graph. xi

$\circ$ entrywise product. 8

col column space. 12

$\square$ Cartesian product of two graphs. 57

$\mathbb{C}$ complex numbers. 1

ℓ_2 ℓ_2 space. 121

$\langle P, Q \rangle$ group or algebra generated by P and Q. 11

$\mathbb{Z}_n$ integers modulo n. 32

$\langle M_i, zz^* \rangle$ inner product of two matrices. 2

$\mathcal{M}$ embedding. 68

$\mathrm{Aut}(\mathbb{Z}_p)$ automorphism group. 99

$\mathbf{1}$ all-ones vector. 3

$\otimes$ Kronecker product of two matrices, or tensor product of two vector spaces. 5

$\mathbb{Q}$ rational numbers. 54

$\sim$ adjacent to. 6

span vector space spanned by. 24

$\mathrm{Sym}(r)$ symmetric group of degree r. 61

tr trace. 42

* conjugate transpose. 1

^{T} transpose of a matrix. 2

e_i characteristic vector of i. 2

dim dimension. 12

gcd greatest common divisor. 31

vec vectorization. 8

References

[1] AHARONOV, D., AMBAINIS, A., KEMPE, J., AND VAZIRANI, U. Quantum walks on graphs. In *Proceedings of the 33rd Annual ACM Symposium on Theory of Computing* (2001), 50–9.

[2] AMBAINIS, A. Quantum walk algorithm for element distinctness. In *Proceedings of the 45th Annual IEEE Symposium on Foundations of Computer Science* (2003), 1–33.

[3] AMBAINIS, A., BACH, E., NAYAK, A., VISHWANATH, A., AND WATROUS, J. One-dimensional quantum walks. In *Proceedings of the 33rd Annual ACM Symposium on Theory of Computing* (2001), 37–49.

[4] AMBAINIS, A., KEMPE, J., AND RIVOSH, A. Coins make quantum walks faster. In *Proceedings of the 16th Annual ACM-SIAM Symposium on Discrete Algorithms* (2005), 1099–1108.

[5] AMBAINIS, A., PORTUGAL, R., AND NAHIMOV, N. Spatial search on grids with minimum memory. *Quantum Information & Computation 15*, 13–14 (2015), 1233–47.

[6] ANGELES-CANUL, R., NORTON, R., OPPERMAN, M., PARIBELLO, C., RUSSELL, M., AND TAMON, C. On quantum perfect state transfer in weighted join graphs. *International Journal of Quantum Information 7*, 8 (Sep. 2009), 1429–45.

[7] ANGELES-CANUL, R., NORTON, R., OPPERMAN, M., PARIBELLO, C., RUSSELL, M., AND TAMON, C. Perfect state transfer, integral circulants and join of graphs. *Quantum Information and Computation 10*, 3–4 (Jul. 2010), 325–42.

[8] BACHMAN, R., FREDETTE, E., FULLER, J., LANDRY, M., OPPERMAN, M., TAMON, C., AND TOLLEFSON, A. Perfect state transfer on quotient graphs. *Quantum Information and Computation 12*, 3–4 (Aug. 2012), 293–313.

[9] BAI, L., ROSSI, L., CUI, L., AND HANCOCK, E. A novel entropy-based graph signature from the average mixing matrix. *Proceedings of the 23rd International Conference on Pattern Recognition* (Dec. 2016), 1339–44.

[10] BARR, K., PROCTOR, T., ALLEN, D., AND KENDON, V. Periodicity and perfect state transfer in quantum walks on variants of cycles. *Quantum Information and Computation 14*, 5–6 (Apr. 2014), 417–38.

[11] BÉNY, C., AND RICHTER, F. Algebraic approach to quantum theory: A finite-dimensional guide. *arXiv:1505.03106* (2018).

[12] BIGGS, N. Automorphisms of imbedded graphs. *Journal of Combinatorial Theory, Series B 11*, 2 (Oct. 1971), 132–8.

[13] BOSE, R., SHRIKHANDE, S., AND SINGHI, N. Edge regular multigraphs and partial geometric designs with an application to the embedding of quasi-regular designs. *Colloquie Internazionale sulle Teorie Combinatorie 1* (1976): 49–81.

[14] BRIDGES, W., AND SHRIKHANDE, M. Special partially balanced incomplete block designs and associated graphs. *Discrete Mathematics 9*, 1 (Jul. 1974), 1–18.

[15] CHEUNG, W., AND GODSIL, C. Perfect state transfer in cubelike graphs. In *Linear Algebra and Its Applications* (2011), 2468–74.

[16] CHILDS, A. Universal computation by quantum walk. *Physical Review Letters 102*, 18 (May 2009), 180501.

[17] CONDER, M., JAJCAY, R., AND TUCKER, T. Regular Cayley maps for finite abelian groups. *Journal of Algebraic Combinatorics 25*, 3 (Apr. 2007), 259–83.

[18] COUTINHO, G. *Quantum State Transfer in Graphs*. PhD thesis, University of Waterloo (2014).

[19] COUTINHO, G., AND GODSIL, C. Perfect state transfer in products and covers of graphs. *Linear and Multilinear Algebra 64*, 2 (Jan. 2015), 1–12.

[20] COUTINHO, G., GODSIL, C., GUO, K., AND VANHOVE, F. Perfect state transfer on distance-regular graphs and association schemes. *Linear Algebra and Its Applications 478* (Jan. 2015), 108–30.

[21] COUTINHO, G., AND LIU, H. No Laplacian perfect state transfer in trees. *SIAM Journal on Discrete Mathematics 29*, 4 (Jan. 2015), 2179–88.

[22] ELLIS-MONAGHAN, J. A., AND MOFFATT, I. *Graphs on Surfaces: Twisted Duality, Polynomials, and Knots*. SpringerBriefs in Mathematics, 2013.

[23] EMMS, D., HANCOCK, E., SEVERINI, S., AND WILSON, R. A matrix representation of graphs and its spectrum as a graph invariant. *Electronic Journal of Combinatorics 13* (May 2005), 1–14.

[24] EMMS, D., SEVERINI, S., WILSON, R., AND HANCOCK, E. Coined quantum walks lift the cospectrality of graphs and trees. *Pattern Recognition 42*, 9 (Sep. 2009), 1988–2002.

[25] FALK, M. Quantum search on the spatial grid. *arXiv:1303.4127* (Mar. 2013).

[26] GODSIL, C. *Algebraic Combinatorics*. Chapman & Hall, 1993.

[27] GODSIL, C. Periodic graphs. *The Electronic Journal of Combinatorics 18*, 1 (Jun. 2011), P23.

[28] GODSIL, C. Average mixing of continuous quantum walks. *Journal of Combinatorial Theory, Series A 120*, 7 (2013), 1649–62.

[29] GODSIL, C. Real state transfer. *arXiv:1710.04042* (2017).

[30] GODSIL, C. Sedentary quantum walks. *arXiv:1710.11192* (Oct. 2017).

[31] GODSIL, C., AND GUO, K. Quantum walks on regular graphs and eigenvalues. *Electronic Journal of Combinatorics 18*, 1 (Nov. 2011), Paper 165.

[32] GODSIL, C., GUO, K., AND MYKLEBUST, T. Quantum walks on generalized quadrangles. *The Electronic Journal of Combinatorics 24*, 4 (Oct. 2017), Paper 4.16.

[33] GODSIL, C., AND HENSEL, A. Distance regular covers of the complete graph. *Journal of Combinatorial Theory, Series B 56*, 2 (Nov. 1992), 205–38.

[34] GODSIL, C., MULLIN, N., AND ROY, A. Uniform mixing and association schemes. *Electronic Journal of Combinatorics 24*, 3 (2017), Paper 3.22.

[35] GODSIL, C., AND ROYLE, G. *Algebraic Graph Theory*. Springer New York, 2001.

[36] GODSIL, C., AND SMITH, J. Strongly cospectral vertices. *arXiv:1709.07975* (Sep. 2017).

[37] GODSIL, C., AND ZHAN, H. Discrete-time quantum walks and graph structures. *Journal of Combinatorial Theory, Series A 167*, 9 (Oct. 2019), 181–212.

[38] GROSS, J., AND TUCKER, T. *Topological Graph Theory*. Dover Publications, 2001.

[39] GUSTIN, W. Orientable embedding of Cayley graphs. *Bulletin of the American Mathematical Society 69*, 2 (Mar. 1963), 272–6.

[40] HATCHER, A. *Algebraic Topology*. Cambridge University Press, 2002.

[41] HEFFTER, L. Ueber Tripelsysteme. *Mathematische Annalen 49*, 1 (Mar. 1897), 101–12.

[42] HORN, R. A., AND JOHNSON, C. R. *Matrix Analysis*. Cambridge University Press, 2012.

[43] KAY, A. Perfect state transfer: Beyond nearest-neighbor couplings. *Physical Review A 73*, 3 (Sep. 2006), 032306.

[44] KAY, A. Basics of perfect communication through quantum networks. *Physical Review A 84*, 2 (Feb. 2011), 022337.

[45] KAYE, P., LAFLAMME, R., AND MOSCA, M. *An Introduction to Quantum Computing*. Oxford University Press, 2007.

[46] KEMPE, J. Quantum random walks hit exponentially faster. *Probability Theory and Related Fields 133*, 2 (May 2002), 215–35.

[47] KENDON, V. Quantum walks on general graphs. *International Journal of Quantum Information 4*, 5 (2006), 791–805.

[48] KENDON, V., AND TAMON, C. Perfect state transfer in quantum walks on graphs. *Quantum Information & Computation 14*, 5–6 (2014), 417–38.

[49] KROVI, H., AND BRUN, T. Hitting time for quantum walks on the hypercube. *Physical Review A 73*, 3 (Mar. 2006), 32341.

[50] KURZYŃSKI, P., AND WÓJCIK, A. Discrete-time quantum walk approach to state transfer. *Physical Review A – Atomic, Molecular, and Optical Physics 83*, 6 (2011), 062315.

[51] LATO, S. *Quantum Walks on Oriented Graphs*. PhD thesis, University of Waterloo, 2019.

[52] LINS, S. Graph-encoded maps. *Journal of Combinatorial Theory, Series B 32*, 2 (Apr. 1982), 171–81.

[53] LOVETT, N., COOPER, S., EVERITT, M., TREVERS, M., AND KENDON, V. Universal quantum computation using the discrete-time quantum walk. *Physical Review A 81*, 4 (Apr. 2010), 042330.

[54] MORADI, M., AND ANNABESTANI, M. Möbius quantum walk. *Journal of Physics A: Mathematical and Theoretical 50*, 50 (Dec. 2017), 505302.

[55] NIELSEN, M., AND CHUANG, I. *Quantum Computation and Quantum Information*. Cambridge University Press, 2010.

[56] PATEL, A., RAGHUNATHAN, K., AND RUNGTA, P. Quantum random walks do not need a coin toss. *Physical Review A 71*, 3 (2005), 032347.

[57] PETROELJE, W. S. *Imbedding Graphs in Pseudosurfaces*. PhD thesis, Western Michigan University, 1971.

[58] PORTUGAL, R. *Quantum Walks and Search Algorithms*. Springer New York, 2013.

[59] PORTUGAL, R., AND SEGAWA, E. Coined quantum walks as quantum Markov chains. *Interdisciplinary Information Sciences, 23*, 1 (2017), 119–25. *arXiv:1612.02448*.

[60] RICHTER, R. B., ŠIRÁN, J., JAJCAY, R., TUCKER, T. W., AND WATKINS, M. E. Cayley maps. *Journal of Combinatorial Theory, Series B 95*, 2 (Nov. 2005), 189–245.

[61] RINGEL, G. Über das Problem der Nachbargebiete auf orientierbaren Flächen. *Abhandlungen aus dem Mathematischen Seminar der Universität Hamburg 25*, 1–2 (Apr. 1961), 105–27.

[62] SADOWSKI, P., MISZCZAK, J. A., OSTASZEWSKI, M., SADOWSKI, P., MISZCZAK, J. A., AND OSTASZEWSKI, M. Lively quantum walks on cycles. *Journal of Physics A: Mathematical and Theoretical 49*, 37 (Dec. 2015), 375302.

[63] ŠTEFAŇÁK, M., AND SKOUPÝ, S. Perfect state transfer by means of discrete-time quantum walk search algorithms on highly symmetric graphs. *Physical Review A 94*, 2 (Aug. 2016), 022301.

[64] ŠTEFAŇÁK, M., AND SKOUPÝ, S. Perfect state transfer by means of discrete-time quantum walk on complete bipartite graphs. *Quantum Information Processing 16*, 3 (Mar. 2017), 1–14.

[65] SZEGEDY, M. Quantum speed-up of Markov chain based algorithms. *45th Annual IEEE Symposium on Foundations of Computer Science* (2004), 32–41.

[66] TERRY, C., WELCH, L., AND YOUNGS, J. The genus of K12s. *Journal of Combinatorial Theory 2*, 1 (Jul. 1967), 43–60.

[67] UNDERWOOD, M., AND FEDER, D. Universal quantum computation by discontinuous quantum walk. *Physical Review A 82*, 4 (Aug. 2010), 042304.

[68] VAN DAM, E., AND SPENCE, E. Combinatorial designs with two singular values I. Uniform multiplicative designs. *Journal of Combinatorial Theory, Series A 107*, 1 (Jul. 2004), 127–42.

[69] VAN DAM, E., AND SPENCE, E. Combinatorial designs with two singular values II. Partial geometric designs. *Linear Algebra and Its Applications 396*, 1 (Feb. 2005), 303–16.

[70] VON NEUMANN, J. Beweis des Ergodensatzes und des H-Theorems. *Zeitschrift fuer Physik 57* (Mar. 1929), 30–70.

[71] WATROUS, J. Quantum simulations of classical random walks and undirected graph connectivity. *Journal of Computer and System Sciences 62*, 2 (Mar 2001), 376–91.

[72] WHITE, A. *Graphs, Groups, and Surfaces*. North-Holland, 1984.

[73] YALÇINKAYA, İ., AND GEDIK, Z. Qubit state transfer via discrete-time quantum walks. *Journal of Physics A: Mathematical and Theoretical 48*, 22 (Jun. 2015), 225302.

[74] YOSHIE, Y. Periodicities of Grover walks on distance-regular graphs. *Graphs and Combinatorics 35*, 6 (Nov. 2019), 1305–21.

[75] ZHAN, H. An infinite family of circulant graphs with perfect state transfer in discrete quantum walks. *Quantum Information Processing 18*, 12 (Dec. 2019), 369.

[76] ZHAN, X., QIN, H., BIAN, Z., LI, J., AND XUE, P. Perfect state transfer and efficient quantum routing: A discrete-time quantum walk approach. *Physical Review A 90*, 1 (May 2014), 012331.

Index

Printed in the United Kingdom by TJ Clays Ltd.